T0295486

A RESONANT ECOLOGY

Richard Misrach, *Cargo Ships (November 20, 2021 6:33am)*, 2022.
© Richard Misrach, courtesy Fraenkel Gallery, San Francisco.

SIGN, STORAGE, TRANSMISSION
A series edited by Jonathan Sterne and Lisa Gitelman

A RESONANT ECOLOGY

MAX RITTS

Duke University Press *Durham and London* 2024

Printed in the United States of America on acid-free paper ∞
Project Editor: Liz Smith
Designed by Courtney Leigh Richardson
Typeset in Warnock Pro and Comma Base by
Westchester Publishing Services

Library of Congress Cataloging-in-Publication Data
Names: Ritts, Max, [date] author.
Title: A resonant ecology / Max Ritts.
Other titles: Sign, storage, transmission.
Description: Durham : Duke University Press, 2024. |
Series: Sign, storage, transmission | Includes bibliographical
references and index.
Identifiers: LCCN 2023053856 (print)
LCCN 2023053857 (ebook)
ISBN 9781478030911 (paperback)
ISBN 9781478026648 (hardcover)
ISBN 9781478059882 (ebook)
Subjects: LCSH: Bioacoustics—Canada, Northern. |
Nature sounds—Canada, Northern. | Marine resources
development—Political aspects—Canada, Northern. |
Marine ecosystem management—Political aspects—Canada,
Northern. | Biotic communities—Canada, Northern. |
Sound—Political aspects—Canada, Northern. | Sound—Social
aspects—Canada, Northern. | BISAC: SCIENCE / Earth
Sciences / Geography | SCIENCE / Acoustics & Sound
Classification: LCC QH510.5 .R58 2024 (print) | LCC QH510.5
(ebook) | DDC 591.59/409719—dc23/eng/20240620
LC record available at https://lccn.loc.gov/2023053856
LC ebook record available at https://lccn.loc.gov/2023053857

To Val Ross (1950–2008)

CONTENTS

ACKNOWLEDGMENTS

My first thanks go to Spencer Greening and Jeremy Pahl: proud Gitk̲’a’ata, brilliant artists, lifelong friends. This book owes a great deal to the examples you set and the experiences you shared with me. To Nicole Robinson and Bunker: thank you for so many visits, for salmon and halibut, and for stories, gossip, and laughter most of all. To Clyde Ridley, one of the great storytellers of the North Coast: “Bat Juice” never tastes the same now, thanks to you. In Hartley Bay, I must also thank many other wonderful people I got to know and work alongside: Johnny Pahl, Priscilla, Mary Danes, Matt Danes, Mona Danes, Marven Robinson, Teri-Jo Robinson, George Fisher, Myron Dundas, Nerissa Bolton, Ethan Dundas, Chris Bolton, Steven Dundas, Toto, Stan Robinson, Gary Robinson, Belle Eaton, Tony Eaton, Ian Eaton, Helen Clifton, Donald Reece. Spending time with you all, fishing, swimming in the creek, losing at basketball, walking to the lake, learning how to chop fish and cut seaweed, drinking endless pots of too-sweet coffee, just spending time in your Territory—these times have enriched my life immeasurably.

Before I even visited Hartley Bay, Janie Wray and Hermann Meuter provided me with an opportunity to learn about the North Coast through a summer internship at Cetacea Lab. Thank you so much. Thank you as well to the community of Gitk̲’a’ata allies who inspired me and made the subsequent work possible: Chris Picard, Kim-Ly Thompson, Eric Keen, Bryn Letham, Hussein Alidina, and Andy Wright. I feel proud to have worked alongside you in support of an Indigenous North Coast. Big thanks to Jessica Rampling (aka Apples): close friend, old bandmate, and fellow traveler across these unceded lands. Thanks to Ken Shaw, Marty Bowles, Mae Jong-Bowles, Carol

Brown, Sarah Chi Brown, Des Nobels, Wendy Brook, Lou Allison, Bill Smith, John Leakey, Doug Bodnar, Ellen Marsh, Maureen Atkinson, Sharon Oskey, Brian Denton, Stan Hutchings, Karen Hansen, Amanda Barney, Jeremy Janz, Tracey Robinson, Jean Eiers-Page (and staff at the Prince Rupert Archives), Carolina de Ryk, Molly Clarkson, and Sarah Stevenson. To all of you, I say this: you may not always agree with everything in this book, but I truly hope that the sincerity of my love for the place you call home shines through here.

I began research for this book at the Department of Geography, University of British Columbia. Here, my deepest thanks go to Trevor Barnes: advisor, editor, and friend. Trevor: I remain so grateful for your patience and generosity. This project also benefited from some outstanding close readers from whom much was asked, and much was given: Jonathan Sterne, Karen Bakker, and Geoff Mann. Karen tragically passed away as this book was being finished. Karen, I owe you such thanks. You were a kindred spirit of eclectic interests; you supported me and gave me opportunities to explore new geographical horizons with you. Jonathan, you were the essential second advisor, giving this project its crucial berth over a semester of sound studies and media studies learning at McGill. Geoff, thank you for the term *sonic materialism* and for your powerful insights and easy generosity—from my first graduate class all the way to my second-to-last beer on Commercial Drive. I must also give thanks to Glen Coulthard, who was a generous critic of my eccentric and sometimes disorderly ideas. My time as a student at UBC Geography was marked by many rich conversations. Thanks to Dan Cohen, Jonathan Peyton, Jon Luedee, Mike Fabris, Dawn Hoogeveen, Andrew Shmuley, Andrea Marston, Jessica Hallenbeck, Matt Dyce, Jessi Lehman, Rosemary Collard, Juliane Collard, Carolyn Prouse, Gerry Pratt, Juanita Sundberg, and Jessica Dempsey for readings, criticisms, and encouragements. Special thanks to Mike Simpson, whose kindness and scholarly example were vital for getting through the hard parts and celebrating the rest.

The School of Geography, Environment and Society at the University of Minnesota is a wonderful place to do postdoctoral work. My biggest thanks here are to Bruce Braun, who oversaw this project's initial transformation with characteristic generosity and insight. At the GES and across the Twin Cities, I had the pleasure of working alongside a wonderful collection of scholars/thinkers: George Henderson, Vinay Gidwani, Arun Saldanha, Gabe Schwartzmann, Mikkel Vad, Sam Gould, and John Kim. Thanks especially to Sumanth Gopinath, who provided crucial insight on this project and whose work exemplifies what a radical sound study can be about. An opportunity to do further postdoctoral study in Sweden came in the form of an invitation to join an interdisciplinary research project, VIVA-PLAN. I wish to thank each of its participants and Natalie Gulsrud for supporting the invitation. The geography community in Lund and Malmö provided wonderful intellectual support when this project seemed to be fading from view. Andrea Iossa, Maria Persdottir, Wim Carton, Gunnar Cerwen, Andreas Malm: thank you for the many walks around what remains, for me, a humbly great example of socialist possibility: Folkets Park. At Cambridge, I refined this text while working alongside a wonderful group of researchers organized by Jennifer Gabrys: Michelle Westerlaken, Danilo Urzedo, Trishant Simlai, and Yvonne Martin-Portugues. Thank you. The incomparable library at King's College was an easy place to revise with the sense of gravitas the revision process sometimes requires. David Good, thank you for supporting me in my role as a College Research Associate at King's. Thanks also to Matt Gandy, Jonny Turnbull, and Elsa Noterman—Cambridge colleagues on Zoom and in spirit, if not always in physical space.

Beyond fortunate meetings at these academic stopovers, I was lucky to develop this book amid other communities of friends and colleagues. Thank you especially to Martin Danyluk, Nathan McClintock, Sarah Wiebe, Sarah de Leeuw, John Shiga, Bram Buscher, Rebecca Rutt, Rafico Ruiz, Gabriel Mindel, Sue Smitten, and the late Stuart Gage for your encouragements. Thank you to Eric Drott, Nick Andermann, Dave Novak, Stefan Helmreich, Alex Loftus, Sara Nelson, Lilian Radovac, and Nicole Starosielski for your generosity as readers and critics over the book's long gestation period. My time at Clark University has been brief, but already I can see the importance of strong collegial support. Thanks to James McCarthy, Rinku Roy-Chowdhury, Gustavo de L. T. Oliveira, Danielle Hanley, and Abby Frazier (surely this list is now expanded). Thanks too to Rowan Compton, who made sure the book's bibliography was in order. At Duke, Courtney Berger was the ideal combination of patience and rigor. The rest of the editorial team—including Laura Jaramillo, Liz Smith, and Aimee Harrison—was outstanding as well (I learned a lot about writing thanks to their collective involvement in this project). An important acknowledgment to the book's funding bodies, which allowed the research to unfold in the timely way it did. This project has received funding from the European Research Council (ERC) under the European Union's Horizon 2020 research and innovation program (grant agreement no. 866006). It also received doctoral and postdoctoral funding awards from the Social Sciences and Humanities Research Council of Canada (SSHRC); King's College (University of Cambridge), and the Willow Grove Foundation. Thanks to Willow Grove, RAVEN, the Royal BC Archives, the CBC offices at Prince Rupert, and the Prince Rupert Archives for sharing research materials and providing general support.

Finally, to Morton Ritts, Maddie Ritts, Zoe Ritts, Dan Rosenbaum, and Sean Yendrys: Thank you for your endless support, laughter, and love. Hans Jacob, you are already surpassing your father in the art of noise. I love you so much. Mischa, you were born just as this book was going to print—I love you so much too! My greatest thanks go to Kelsey Johnson. You know more than anyone how lucky I am. For this project and for so much more, you are the real MVP.

INTRODUCTION ON A RESONANT NORTH COAST

In December 2011, I traveled to the North Coast to spend a rainy week listening to whales at a makeshift research station called Cetacea Lab, on the southern tip of Gil Island. Headphones on and left alone most of the time, I considered the recordings of humpback whales the lab had taken the previous summer. These were, cofounder Janie Wray explained, songs, and they communicated the returns of a species that had been extirpated from the region for decades. I sat there, letting the songs come at me in trumpetings, pitch shifts, and intensity bursts; the hydrophones that recorded them hissing steadily beneath. They were profusely beautiful sounds, but it was their geographical qualities, not their musical ones, that had inspired my visit. In the early 2000s, this remote section of northern British Columbia became marked by an unsettling concurrence of events: humpback whales reappearing at almost the exact same time the Canadian government began broadcasting its support for regional energy development, including a multibillion-dollar shipping project (Enbridge's Northern Gateway project), which would bisect many marine habitats.[1] At Cetacea Lab, songs were being recorded to support conservation efforts that would protect whales and other species from Enbridge and the existential risks it posed. The extraordinary nature of the effort had not gone unnoticed. Interest in Cetacea Lab would grow so rapidly in those years that Janie and Hermann Meuter (the lab's other founder) felt compelled to develop a volunteer program, "to help us manage the whale recordings and all the visitors wanting to come hear them," as Hermann put it. By 2011, Cetacea Lab was hosting collaborations with documentary filmmakers, ecotourists, and philanthropists, as well as a multitude of scientific research teams. New activity at Whale Point was impressed upon me in each of the summers I visited (2012–15, 2018). It was easy to grasp why Cetacea Lab's work—with its array of listening technologies, suspended in the quiet waters of Taylor Bight, Squally, and Whale Channel—could produce such abundant evidence and why, with its mist-soaked rainforest and mountain backdrop, environmental groups would so eagerly look to feature it in their campaigns. At the same time, I would become preoccupied with the idea that Hermann and Janie were taking something else from the songs. More than a beautiful sound or piece of evidence, whale song was also, for them, an artifact of loss; a loss expressing hugely historical ideas of humanity, nature, and the relations between.

Cetacea Lab was built on the unceded lands of the Gitga'at First Nation. Since 2001, it has operated on the permission of the nation's traditional and elected leadership. For years, Hermann and Janie had little contact with the community. It was only with the rise of Enbridge that dialogues with the Gitga'at village of Hartley Bay (pop. 120) became routine. Hartley Bay is the closest human settlement to Cetacea Lab, at an hour's boat ride. By the time of my visit, it was also ground zero for anti-Enbridge sentiment up and down the coast.[2] For the big environmental NGO reps who had descended from Toronto, Vancouver, and further afield, celebrating the Gitga'at Territory's spectacular natures—sharing them across webpages and films—was the best way to protect them. This was not without some proof of concept: it chimed with the earlier forms of media-savvy conservation that have made Canada's West Coast a beacon of environmentalism for decades.[3] But not everyone in Hartley Bay was convinced. One dissenter was a young Gitk̲'a'ata hunter named Spencer Greening. Spencer had spent his teenage years fusing the extreme sounds of black metal music with the traditional drumming of his Ts'msyen culture.[4] When I met Spencer, he was twenty-three, and on the cusp of assuming a leadership role with the village band council. Spencer loathed Enbridge, but he also opposed the way "environmentalists like to turn our territory into a theme park." He had formed a band, Gyibaaw, with his cousin, singer and guitarist Jeremy Pahl, several years before. I was able to find a cassette of their debut album, *Ancestral War Hymns* (2009), by hunting around black metal pages on Reddit. Following Jeremy's advice, I put it on one evening while driving along the Skeena—the huge river that forms the cradle of Ts'msyen culture.[5] It was unquestionably abrasive, but as I listened, its cacophony of screams, distortion, and kick drums began to synchronize with the water flowing at my side; the moonlit movements of the river resonating with the harsh repetitions of the music. I have listened to the album a hundred times since. Each time, I am struck by the strange timeliness of its musical vision:

an Indigenous North Coast shorn of capitalist and colonial rule, in which an assortment of peoples and animals, waters, laws, and spirits all move in a complex unity.

This book is about a geography apprehended through sound, made through sound: a resonant ecology. It is dedicated to the great sonic visionaries of the North Coast: the whales and Indigenous musicians who affirm the region as an incomparable nexus of place soundings. The North Coast in the first two decades of the twenty-first century produced a multitude of sounds inscribing an audible development process. Humpback song and Indigenous metal echoed out alongside a host of other soundings, some dissident, and some assenting; some familiar, and some inscribed with global transformations. Their polyphony, expressive of the buildup phase of a massive industrial expansion, reveals sound as a powerful site for charting how nature-society relationships are made and unmade in geographies marked by capitalist and colonial change.[6]

Critical engagements with the ecological dimensions of sound are hardly new. But the proliferation of recent interest here—from the extensive theoretical work on sound, to the celebrated findings of bioacoustics and ecoacoustics, to the civic promotions of citizen sensors—is significant.[7] Looming large are the twin specters of the Anthropocene and climate change, whose audible expressions reach from the tremors of calving glaciers to the high chirps of invasive insects. Through their combination, questions of sound have come to incite new speculations on nature and agency, and the prospect of reawakening to forces that have long been in humanity's collective midst. For some thinkers trying to perceive the world in new ways, modernist constructions premised on distinctions between humanity and nature have been exposed as outmoded. Climate change has ended the assumption of a stable setting upon which human activities take place. "In an age of global warming," writes Timothy Morton, "there is no background, and thus there is no foreground."[8] What there is, for Morton, is a destabilization of material forms and the analytical frameworks designed to address them; an undoing of the sensory hierarchy long designated by modernity.[9] Marx claimed that the "formation of the five senses is the labor of the history of the world."[10] He linked the question of humanity's sensate relation to nature to the question of humanity's encounter with capitalism, out of which the senses develop as historical products.[11] But what about the role of other forces in the emergence of human sensibilities? What about the other life-forms being resensitized in the age of the Anthropocene? How might we narrate and seek to understand their stories? Perhaps radical new frameworks are needed for making sense of the "instances of recognition" being forced upon us today.[12]

The foregoing has become a prevailing wisdom in many ecologically inclined disciplines, including the one in which I was trained (geography). And while there is much to recommend it, such as the call for greater attention to the nonhuman, it tends to come with a curious assumption: the notion that recent turns to sound are inherently progressive in nature and, if properly socialized, will only aid in the realization of the ecological consciousness needed today.[13] This assumption is problematic. Routinely unaddressed in Morton and the other exponents of what we might call the Sonic Anthropocene are enabling developments, such as the proliferating economic interests I discuss in this book. These developments pose additional questions. To what degree are ethical frameworks primed to ideas of listening ignoring the ways powerful corporations and the surveillant state are also moving in this direction?[14] How might the popular scientific claim that "nature is always listening" describe evolving relations of nature and society, not the obsolescence of these categories?[15] And not least, what sorts of transformative pathways might some of these sonic turns actually be working to inhibit, and at a time when many of the communities most attuned to difference are suffering escalating forms of loss and violence?[16]

In this book, I approach sound as a constitutively indeterminate figure in the politics of nature, a social object whose relations and attributed meanings are contingent outcomes of webs of mediation—biophysical, economic, cultural, socio-technical. As a work of sound studies, *A Resonant Ecology* is also a work of historical materialism. It seeks to continue historical materialism's inquiry into the ways the natural and the social are locked in a "dialectic whose boundaries are to be determined," while taking on new insights regarding agency, place, and science.[17] It considers how new collective interests in the interstices of environments—their insides and outsides; their enabling elements; their pasts, presents, and futures—inform sonic acts (e.g., listening, recording, composing). Through a situated ethnographic engagement, it asks how different communities—group-differentiated, collectively realized engagements with sound/music/noise—know, manage, and contest spaces being programmed for industrial change. Acoustic monitoring at a whale research station; measurements of ocean noise in an international scientific research project; ecological place attachments in an old fishing village; compositions of Indigenous black metal; smart eco-governance projects—each points to enactments involving different kinds of sound. Through them, this book offers a sonic materialist analysis of contemporary environmental politics; a critical account of development and transformation on the North Coast.[18]

Sonic Materialism, Mediation, and Limits

Two observations guide this book's arguments. The first is that a distinctive set of turns toward sound would take place along the North Coast during the years featured here (roughly 2006–18). Sound often gives form to "what is steadily marginalized or located within the more peripheral zones of presence," as Brandon LaBelle suggests.[19] There is a wealth of critical geographical work documenting how sound can extend opportunities for tracing coproductions of space, place, and nature.[20] These studies cover everything from multispecies ethics to the socio-technics of urban policing. That a single sound—the ringing of a bell, the broadcasting of a voice—can be said to express a multiplicity of relations might seem a difficult claim to accept. But a central aim of this book is to show through a historically attentive sensibility how sound's transitory natures can unfold the deep structuration and entanglements of the human and natural world. Sonic productions can belie highly conditioned arrangements of material and semiotic elements, replete with theories of listening and cultures of response. They can point to new conceptions of time and space. Above all, the sounds I consider here exist as stories of the changing North Coast. Rather than contingent events, they find their effect within a geography coproduced at multiple scales as a resonant ecology.

What does this mean? In the first, it means a sonic materialism capable of acknowledging the North Coast as much more than a backdrop to this study. Rather, the North Coast is a space marked by different kinds of sonic agency, encompassing humans, animals, and the distinctive capacities of places.[21] And moving among these is a new cohering force: development and its coming boom. Capitalist development is always a disruptive force, both a promise and an imposition. What concerns me in the present instance is the search for new points of resonance it will occasion: the how and the why and the through what means a massive expansion of industrial, logistical, and merchant capital is to be made sense of within an already complex region. On the North Coast, political economies of development are sanctioning new sensitivities to nature in the form of acoustical monitoring observatories, experimental biosciences, changing urban/rural place attachments, shifting regulatory regimes, and the tuning and management of affects. Local turns to sound are intersecting with state surveillance projects and regulatory regimes and spurring a range of counterlistenings and composings. Amid the activities I have witnessed, I have encountered new convictions about what sound is and can be about. They involve new subjectivities and material situations. Some are guided by particularistic local concerns, many are consistent with the universalistic, white, bourgeois figure of Jonathan Sterne's "Audio-Visual Litany," and still others invite us to consider nonhuman and multiply-agentic coproductions of nature.[22]

A second guiding observation of this book concerns digital technologies. The same period that witnessed a diversity of local turns also witnessed a decisive transformation in sound culture at the global scale: digitalization. Sound's becoming discrete, as when waveforms become sampled and thus partitioned at set intervals, has had transformative effects on development geographies worldwide. While the history of sound's digitalization goes back decades, the rapidly developing industrial armature that would begin to shape this process beginning in the early 2000s (from file-sharing protocol and recorder miniaturization to software, algorithms, and sensors) has meant dramatic changes in sound's capacity to direct and inform environmental governance. Karen Bakker refers to the new epistemological possibilities being opened through digitalization as a revolution in "sonics"—akin to the optics that guided the scientific revolution.[23] The proposition of sound as a kind of freely available resource has gained a wide institutional uptake: raw material for extractive industries, digital arts, surveillant states, and innovative technosciences alike. There are powerful logics at work here, and too many lineaments to summarize. But a unifying ethos, of particular significance to this study, appears in a quote from an unnamed source working with IBM's Acoustic Insights Program in 2020: "sound has so much hidden data."[24]

This idea of sound qua data, with data being the central commodity of digital capitalism, has been brilliantly explored in the realm of music by Eric Drott. In *Streaming Music, Streaming Capital,* Drott shows how music's observable capacity to produce data about intimate sites and spaces of social reproduction has become monetized by music streaming platforms and turned into a new tool of mass surveillance. But digital sound is also a fecund site for producing socially useful material about myriad environmental changes. In a North Coast where Smart Oceans Systems are expanding capacities for predictive analysis, animal tracking, and real-time regulation along heavily shipped trade corridors (chapter 5), networked digital technologies are apprehending underwater sound as a resource for sustainable marine development. In turn, they are rearticulating the ocean medium in potent and pointed ways. This is not a smooth displacement of the old for the new—a fiction speaking to the very ideology smartness means to impose. Rather, as I stress at several moments in this book (e.g., chapter 1), sonic claims and practices being established through digital technologies are intersecting along the North Coast with other valorizations and lifeways, producing a range of negotiated experiments. Alongside the emergence of new sensor technologies and the technologically mediated search for new animal intelligences are more mundane reorientations: musical integrations of community, individual presentiments of loss. Instead of clean breaks between "sonic epistemes," digitalization is producing new admixtures and gaps; ideations we can hear in a resonant ecology and use to compose that ecology analytically.[25]

To study these dynamics, this book employs a method I call *sonic materialism.* A dialoguing across several scholarly traditions—above all, historical materialist geography and critical Indigenous studies—sonic materialism is guided by the observation that sound exists in and through its complex and shifting mediations. By *mediation,* I refer to "activity which directly expresses otherwise unexpressed relations."[26] Mediation implies the world-disclosing powers of media, whose technological forms (e.g., hydrophones, guitars, maps) can transmit both content and associated cultural meanings. This book seeks to account for the different ways sonic mediations (and media) compose the North Coast and how interlinked contests over sound's measure, meaning, value, and effect can index broader forms of struggle. In this, I draw from Theodor Adorno, who sought to explore how the mediations and media of music—from the classical music of Beethoven to the hiss of the record player on which Beethoven is played—are expressive sites of historical transformation.[27] Adorno also helps elucidate sonic materialism's structural critique: to make sense of the North Coast at a time of uneven development; to trace in sound the mediations that connect a panoply of elements in order to identify the conditions coproducing them. Alongside the dialectical model, this project also advances assemblage approaches to sonic mediation. In assemblage thinking, Julie Guthman writes, "nonhumans play an active role in bringing phenomena into being."[28] Assemblages reveal how particular combinations of mediations can characterize sonic productions. This focus on distributed objects and nonhuman forces can reveal how a range of elements can constitute sonic mediations, in ways that exceed the stipulations of Adorno's model.

As such, the central object of investigation in this book is a kind of multiplicity. Sound is a contested object formed through the mediation of social and material forces that are expressed and (variously) legible in its internal form. Sound can also operate as one of a group of agents articulated in an assemblage. At still other moments, sound can emerge to orchestrate and legitimate already entrenched inequalities of power. So how do we arbitrate between these tendencies and the various realities (social, political, technological, spiritual) they denote? My answer here is to return to the normative impulse of historical materialism. For David Harvey, historical materialism entails "a study of the active construction and transformation of material environments (both physical and social) together with critical reflection on the production and use of geographical knowledge within the context of that activity."[29] In a similar vein, my aim is not to catalog all the mediations at play at a given moment. Rather, it is to pursue those forms necessary to the book's critical project: a situated critique of capitalist-colonial development.

A sonic materialism defined as such is not an innocent critique. It has as its inheritance a Marxist tradition marked by the Enlightenment pursuit of knowledge; a tradition that was forged in the belief that ferreting out the unseen forces capitalism inserts into the world can be a basis for capitalism's critique.[30] The non-innocence of this idea is clarified by considering its Eurocentrism. Insofar as questions of access, transparency, and ethics mark investigations of sound on the North Coast, any account seeking the term *critical* must acknowledge this intellectual provenance. On the North Coast, this means confronting the colonial legacies that continue to shape perceptions of the region and its inhabitants.[31] Dylan Robinson's critique of possessive listening practices, as developed in his book *Hungry Listening*, is helpful here. Robinson foregrounds the central importance of asking for whom a sound is made meaningful and why. As Vanessa Watts, Kim TallBear, and Leanne Simpson further demonstrate, categories like knowledge and justice are not derived from universal doctrines and abstract principles. Rather, they are keenly situated, informed by protocol, and involve a host of human and nonhuman participants. This is a significant observation for a region from which anthropological practices of salvage would gain their wide disciplinary and even cultural assent. It was in the lands and waters of the North Coast that Franz Boas, salvage's great exponent, arguably inaugurated the discipline of sound studies with his efforts to classify the "alternating sounds" of disappearing Ts'msyen speakers.[32] It bears emphasizing that his prefigurative sonic extractions were about more than academic knowledge production. They worked to inform projects of Anglo-European modernity and Canadian state conquest; projects that in various ways sought to incorporate the North Coast and its peoples under banners of progress, improvement, technological rationality, or simply white supremacy.

The analytical and ethical perils of what amounts to a white settler's fieldwork on Indigenous lands and waters (e.g., my own) points to the critical importance of limits in this account. To explain, I need to unpack this polyvalent term. Political ecologists approach limits as "differentially malleable conditions of possibility for different forms of human activity."[33] They suggest capitalist development as routinely a story of the overcoming of limits, given capitalism's endless interest in converting limits (of production, exchange, etc.) into "mere barriers."[34] For Gavin Steingo and Jim Sykes, sound studies can also be described in terms of limits—in this case, as "an ethnographic experiment" with culturally inscribed forms of audibility.[35] Both formulations hold insight for this project, but the limits that most interest me involve something else: responsible research in Indigenous territories. For Audra Simpson, acknowledging limits can mark an awareness of a sovereign order. Simpson develops this idea in her powerful critique of ethnographic accountabilities, *Mohawk Interruptus*. "Rather than stops, or impediments to knowing," she writes, "limits may be expansive in their ethnographic nonrendering and in what they do not tell us."[36]

It is worth pausing to consider the implications of this idea. Simpson is not saying that once the ethnographer encounters limits along "Research Path A" they should go about their way along "Path B," armed with the productive knowledge that "Path A" was unavailable. Rather, limits are better understood as being about a community's right to self-representation. They are about researcher commitments to this right, to community well-being and the maintenance of relationships among humans and the natural world. From the nineteenth century to the present day, sounds have been extracted from the North Coast and made into the possession of others—as culturally valued music, as quantifiable animal calls, as spectacle, as data. Against this possession, sonic materialism must pursue "counter-possession"—radically refusing both "hungry listening" and the unjust conditions it sanctions.[37] If this moment can help to establish a normative, ultimately decolonial project from this study, counter-possession is also useful for the meta-reflection it invites of the book more generally: What are the limits of sound to the study of the North Coast, to its unassailable rhythms and shifting temporalities, to the Indigenous actualities it effectuates in people like Jeremy and Spencer?

I still need to identify the other element in sonic materialism. This is its flipside: sonic capital. Sonic capital is a latent element in this account; the actor waiting offstage, hovering as potential.[38] But sonic capital is essential to recognize if we are to grasp the political stakes this project raises. On the North Coast, between 2006 and 2018, different communities will provide routes through which capital will seek to move, and through which new opportunities for capitalist valorization will arise. Collective capacities realized in the wake of new environmental challenges, socio-technical developments, economic conditions, and regulatory adjustments will become opportunities for, among other things, sonic capital: sonic inputs into the capitalist valorization process.[39] Sonic capital is not the sounds of capitalism—as might be discerned, for instance, in the ocean-bearing noise of global maritime traffic. Rather, to continue the example, sonic capital is what emerges when technoscience works to apprehend ocean noise as a fungible risk/opportunity (chapter 2). Sonic capital *becomes* in the assetization process; in the abstraction of whale listening into exchangeable (and profitable) sense data—activity that is presently spurring lucrative markets in industrial noise assessment and data analytics.

A clear place we can observe these logics is in the construction of Smart Oceans (chapter 5) and the way sonic data is being enrolled in the creation of coastal listening stations (sonic fixed capital). The ambient sound plots Smart Oceans generates to predict marine flows, like the animal recordings acoustical analysts are leveraging to improve data products, reveal noise as a consistent figure for sonic capital generation.[40] The noise/value relation has been an interest for sound studies since the work of Jacques Attali, if not earlier.[41] It proposes numerous links to the present study. To listen to Gyibaaw's noisy music (chapter 5) is to ask after capital's perpetual dependence on "free gifts" of human and nonhuman creativity, and how noise becomes cultural commodity. To listen to Gyibaaw is also to ask how cultural forms might resist sonic capital and its depredations. There are deeper dynamics at play here. The fact that epistemologies marked by partial perspectives, plural agencies, and recursive temporalities are appearing with equal force across decolonial Indigenous art and algorithmically managed smart environments is not a fact we should overlook. It speaks to the Janus-faced character of today's resonant ecologies and the need to employ tools capable of attending to the varied ways contemporary ideas of sound can articulate with material life and economic possibility.

Sonic capital is not preeminent among the different circuits of capital now reshaping the North Coast. And its prospect should not absolve us of the need to attend to those other axes of social domination at play, such as neoliberalism and settler colonial rule. But sonic capital helps to focus a key question that spans the different cases of this book: What kind of a North Coast does capital desire? The answer, I will wager, is this: a predictable one. A predictable North Coast is a coast readied to absorb new forms of transnational marine activity (e.g., shipping). It is a coast whose rhythms can be made manageable, whose anomalies can be monitored and mitigated before they occur, and whose different imaginaries can be guided into the form of a "sustainable marine development."[42] This is the horizon to which sonic capital—and capitalism more generally—strives in the present context. It is to this end, with its contradictory demands of limitless and sustainable growth, its false synthesis of capitalism and ecology, that Smart Oceans technologies are presently listening to Chatham Sound, the Douglas Channel, and other to-be-shipped areas of the North Coast. And to which several other sonic practices, discussed in this book, also press. There is much to question about this vision, including its evacuation of "differences, temporalities, and societal structures."[43] But if this book has an argument in support of sonic materialism, it is to be found here too.

The North Coast as a Development Geography

Liminal geographical zones of capitalist-colonial activity are often marked by overlapping cultures of sounding and sense-making.[44] This is indeed the case for the region in question here: a rugged collection of islands, continental shelf, and sparsely populated coastal fjord land extending over 800 kilometers from the tip of Vancouver Island to the bottom of Alaska. The North Coast has known many names over recorded history, with the actual term *North Coast* only being consecrated in 1990 (as a provincial electoral district for the Legislative Assembly of British Columbia, Canada).[45] As many written accounts about it emphasize, it is above all a watery space, dominated by huge ocean swells, moody gray skies, and seemingly endless rain. Water assumes many functions here (as it will in this book). The marine biologist E. F. Ricketts referenced water in his depiction of the North Coast as an interplay of "quiet water" and "wave-shock" life-forms.[46] Industry boosters have long emphasized the advantages of the region's maritime trade routes—linking North America to the Asian metropoles more speedily than anywhere else. For the Ts'msyen, water is the very medium of life: an ocean (*gyiyaaks*) unfolding in rising (*leeksa'aks*) and falling (*tsoo'aks*) tides. Water is the etymological source for *sounding* and *noise*, concepts that assume shifting local meanings.[47] Rather than just an elemental condition, it is helpful to conceive of water as a "hydro-medium," a condition that will enhance and extend sound's properties and political lives in various observable ways.[48]

The North Coast I would come to know through watery and terrestrial sounds is also an Indigenous North Coast: a space of stories and territories, defined through the circulations of the Ts'msyen who have occupied it for thousands of years: Gitga'at, Gitxaala, Metlakatla, Kitasoo/Xai'xais, Lax Kw'alaams, Gitsumkalum, and Gits'ilaasü—circulations shared with Gitksan, Nisga'a, Tlingit, Haida, and Haisla.[49] Recent years have brought renewed public attention to these histories, including in ways that have benefited Indigenous peoples. But it would be a mistake to minimize the countervailing reality the development process serves to highlight. For most visitors, the remote, storm-battered North Coast remains a "purification machine"—a place where history is to be left behind.[50] In Jean Barman's words, the North Coast is "the West beyond the West," the temporalized edge of the projective, expansive frontier central to settler colonial ontology.[51] Ferrying up the Inside Passage from the northern tip of Vancouver Island, I can still recall how the North Coast first appeared to me: open waterway after open waterway, buttressed by huge walls of hemlock and cedar. It suggested a great indifference to human scale and activity, an impression I shared with the person sitting next to me, an elder from the Heiltsuk village of Bella Bella. He smiled and called me *k'amksiwah*. As we sailed on, he pointed out patterns of second-growth forest, lightly etched lines of logging roads and power lines where I had seen only brush. He noted locations where gold mining was still taking place, behind hillsides. There was more, he added, but he was feeling too tired to talk further.

Since the early 2000s, environmentalist narratives of wilderness have increasingly aligned with scientifically informed appraisals of the North Coast as a biodiversity hotspot. This framing has been incredibly productive for the storied encounters with whales, spirit bears, and coastal wolves the region uniquely provides.[52] Key to this framing were the Great Bear Rainforest agreements of 2004–6, which secured new financing for local forms of economic development and ecosystem-based management. Coastal First Nations—a political alliance made up of nine nations, including the Gitga'at—would use the new resources to advance collective frameworks of Indigenous regional stewardship. One result was the Coastal Guardian Watchmen Program (founded in 2006): "the eyes and ears of the Land and Sea."[53] The Watchmen are central to many of the governance successes that have transpired in the region since that time. Certain members would also become key interlocutors to my own regional understandings, as I relate below.

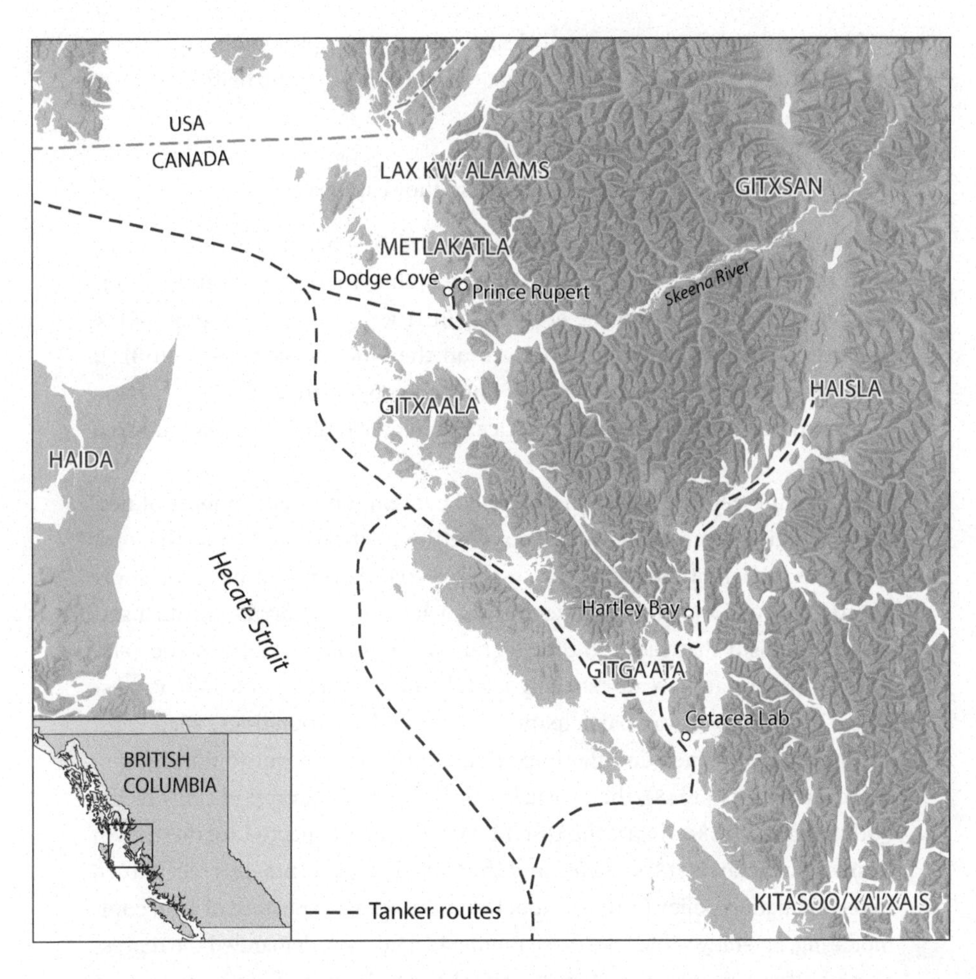

MAP I.1. The North Coast.

Initiatives like the Guardians are hopeful endeavors and cut against a dominant logic touched upon already: the land-centered project of settler colonialism. In this book, I propose that settler colonialism is routinely a "sound-centered" project too.[54] Early in my research, I found a story in Marius Barbeau's 1951 book *Pathfinders of the North Pacific* about Metlakatla—a Ts'msyen village just across the harbor from Prince Rupert. It concerns an encounter between William Duncan, who was a nineteenth-century Anglican priest, and an unnamed messenger from a local First Nation. At issue is Duncan's unwillingness to cease ringing the newly installed village church bell, despite requests:

"The bell so disturbs our mysteries," the messenger pleaded.

"Would you not be kind enough, today, not to ring the bell?"

"I can't do that."

"You could ring it softly?"

"No, I will have to ring it as usual, so they can hear it."

The story is likely apocryphal. The only record Barbeau had access to came from the self-aggrandizing Duncan, and via John Arctander's hastily assembled 1909 biography.[55] But the episode is instructive for its evocation of the despotic character of Duncan's tenure on the North Coast (1857–1918). It also speaks to the sonic colonialities Duncan supported, and which remain central to grasping the region's political character today, over one hundred years later.[56]

From nineteenth-century surveyors to contemporary enactments of networked digital monitoring, Indigenous-settler relations on the North Coast express a recurrent aural facticity: a settler colonialism that has repeatedly sanctioned new and expanded forms of social and territorial control; a settler colonialism that seeks to disintegrate prior social formations and relations to nature and create "socially useful" ones. More than a solitary gesture, Duncan's bell was participant to a network of soundings, silencings, and listenings. It included the impositions of Christian music during missionization (1857–1948), the censuring of Indigenous aural traditions in the Potlatch Bans (1884–1951), the disciplinary silencing enacted by the Indian Residential Schools (1879–1948), and the Canadian government's recent turn to reconciliation policy (2015–), modeled in the reflexive posture of an accommodating, listening state. Like these moments, Duncan's demanded the repression of other sensings and other temporalities.[57] As an audible formation on the North Coast, his bell prefigures an increasingly capacious colonialism too: now underwater, in the air, and amid circulating digital sound objects.

This book's critique is not with colonialism, however, but capitalist colonialism. More needs to be said about interaction of these concepts, and the ways they combine through development. Many geographers have observed how settler colonial regimes do not function as singular logics but are instead coproduced in relation to capitalism's global political economy.[58] The five cases explored in this book share origins in the early 2000s, a period of sustained efforts by the Canadian state to support new global production networks across the Pacific Rim.[59] The central role the North Coast might play in Asia-Pacific integration has long been the stuff of local lore—in high school, children in Prince Rupert are regularly reminded that the trade routes connecting their city to the Asian metropoles are a day's voyage closer than from any other North American port.[60]

In step with China's ascendance as an industrial superpower with unprecedented energy needs and faced with a continental glut of fossil fuels, pressures to forge new supply chain infrastructures in North America began intensifying post-millennium. Among other places, they eventually found expression in the eastern and coastal portions of northern British Columbia. Here, new state provisions would be established for oil and gas extraction as well as bulk container cargo shipping activities. Port expansion projects were pitched in Prince Rupert and Kitimat, along with infrastructure upgrades to regional rail and road networks. These were followed by rounds of digitalization, linked to governance but also to narratives of economic well-being. Throughout the region's inner fjord land and across its outer shelf, the federal and provincial governments conceived of high-speed internet networks capable of supporting new programs of oceanographic profiling, mapping, and tracking.[61] Eventually, large-scale marine planning projects like the World-Class Tanker Safety System (2014–17) would evidence the considerable investments Canada made to ensure the economic transformations that met its policy objective of sustainable marine development. In Prince Rupert, a powerful boosterism took hold, with community newspapers and City Hall making promises and threats not to stand in the way of progress.

Environmental governance is a crucial mediator between capitalist colonial logics and the biophysical conditions they encounter. Following Gavin Bridge and Tom Perreault, this book conceives of environmental governance as a decentralized process, encompassing state interests but moving beyond the state's remit.[62] As suggested earlier, from the mid-1800s to the present day, the thrust of environmental governance on the North Coast has been to maintain orderings of colonial law and support the capital accumulation process. But in recent decades, new political actors—bearing new forms of transnational science, social organizing, and legal authority—have begun to exert influence over this trajectory.[63] Through its focus on sound, this book considers some of the novel engagements that have resulted—from within contested measurements of ocean noise, where data analysts seek to cost the effects of shipping by assessing the sensitivities of constitutionally protected marine species, to emergent forms of Indigenous music, whose "performance of other worlds" articulates new community demands.[64] As nonhuman participants to both processes, whales deserve special mention here. Among other things, whales are supremely acoustic creatures, with distinctive cultures of musical composition and acoustic-spatial awareness that give them elevated positions of authority and respect in many human cultures. But whales are also among the most resourced animals in the annals of western civilization, a historical fact that encompasses over four decades of industrial whaling along the North Coast. As I note in chapter 1, these trajectories co-constitute a taxon that is both cherished as an object of eco-politics and exploitable as a resource in the same sounded moment. In industrialized waterways around the world, whales are straining to listen past raucous shipping conditions while being pursued by conservation drones and acoustic monitoring technologies.[65] Violent new behaviors are being registered, suggestive of a growing metabolic rift that manifests formally in ocean noise. But this is not the only story in which whales feature. As I relate in chapter 1, unexpected flourishings are also happening for certain species on the North Coast. We can find hope in institutional efforts that affirm different kinds of human-whale relating, that work to value nonhuman nature and bear witness to it in remarkable ways too.

My principal fieldwork site of Prince Rupert (pop. 11,500) offered a special vantage onto the changing region. The hub for the region's network of fishing villages and work camps, Prince Rupert profiles as a classic depressed resource periphery, with a Main Street expanse of abandoned buildings and little-used sidewalks. It was founded in 1910 as the terminus of the Grand Trunk Railroad, when it was enshrined with the promise of Asia-Pacific economic integration that has marked the North Coast over the ensuing century plus.[66] But behind this staid appearance are undercurrents of surprising depth, richness, and complexity. Prince Rupert has the largest Indigenous population per capita of any city in Canada, and despite decades of institutional racism and high rates of poverty, numerous Indigenous cultures flourish there. Alongside the settler fishing and logging communities who arrived in the mid-twentieth century, and the older waves of Italian and Norwegian immigrants, the city boasts more recent communities of Sikh missionaries, Vietnamese crabbers, and Filipino dockworkers. They can all be found downtown, conversing in that great polyglot space of the Canadian North: Tim Hortons. Many of the materials that inform this book emerged from these community spaces. They afforded me opportunities to consider how locally distributed references to sound—in personal anecdotes, newspaper editorials, or industry pamphlets—could index the development process.[67] The resulting methodology is experimental in that it recursively moves across different sites, scales, and methods—from archives, regulatory review, and structured interviews to sound walks, boat trips, and personalized headphone journeys—to build an ethnographic account of the different communities operating in the region.

By the end of my first spring, I had acquired a sense of the general reactions development prospects had elicited across different sectors of the region. Environmentalists despaired of the contamination risks and impacts on wildlife. Municipal and city officials pointed to budgetary shortfalls amid a growing housing crisis. Fishermen in Oona River and Dodge Cove worried about further attrition to already diminished economies. But many working-age residents in Prince Rupert eagerly awaited development. It promised an end to high regional unemployment and youth out-migration.[68] Here, two decades of neoliberal policy have all but destroyed the once-vaunted civil society institutions and a city's status as a union town. While the initial play of the Enbridge Northern Gateway project fell flat ("If anything," Hermann Meuter told me, "Enbridge has only further united this coast against development"), liquefied natural gas (LNG) proposals would prove far more considered. As journalistic reportage later confirmed, LNG proponents were able to skillfully exploit local economic anxieties while assuaging ecological concerns through the presentation of an innocuous fossil fuel resource.[69] LNG drew attention away from the huge upgrades ongoing at the Port of Prince Rupert's Fairview Terminal, another key facet of local development ambitions. The early 2000s saw the arrival of new road, rail, and port infrastructures, along with programs for job reskilling and training. Although largely greeted as apolitical, the ongoing penetration of digital connectivity into these spaces would also begin to reshape politics in consequential ways. In the form of smart technologies, digital computation would become a key ideological support for extensive state investments in coastal monitoring and emergency response. Taken as a whole, the result of these interlinked developments (development boosterism, policy neoliberalism, digitalization) has been a local population primed to abandon historical identifications with fishing and logging and pressed into accepting a brave new world of network-enhanced supply chain movement.

The politics of development become murkier when we move to the question of Indigenous communities. Because the lands and waters encompassing the North Coast, like most of the province of British Columbia, were never settled through historic treaty or surrender—that is, the conditions for legal Crown possession under the terms of the Royal Proclamation (1763)—key legal questions surrounding territorial access remain unresolved. Ensuing state-led efforts to pacify and preempt Indigenous oppositions to development—whether through patronage, assurances, threats, or intimidation—have produced mixed results. Decades of impoverishment and inadequate state services have no doubt compelled some communities into seeing recent energy prospects as an opportunity to gain some measure of economic security. But while impact benefit and sharing agreements would be signed between energy proponents and several local band councils, development would also mobilize powerful assertions of Indigenous refusal and place attachment. This is exemplified in the dramatic appearance of Indigenous water and land defenses at energy choke points along the coast and inland. Some of these continue to gather international attention as this book goes to print.[70] Meanwhile, hereditary and elected First Nations leaders armed with important victories in the Canadian courts would build on earlier efforts to reshape resource co-management regimes. This has boosted local capacity in areas of conservation and marine stewardship and inspired innovative forms of intercultural science and research. The socioeconomic changes have been less salutary. Ten years after I first moved to Prince Rupert (2013), it is the urban Indigenous communities who continue to suffer most from the destabilizing effects of development, as measured in high rates of houselessness, poverty, and substance abuse.

This, then, is the historical-geographical arc surrounding my time on the North Coast: a time of uneven development, political unrest, and rapid technological change. A time of the Anthropocene, when algae blooms, red tides, and surging storm systems entered more deeply into a region's lifeways, along with new migratory bird populations, unexpected sightings of Humboldt squid, and movements of coastal wolves. In the following chapters, I consider some moments that emerged within and alongside these changes. I reflect upon the analytic powers of sound and sound's analytical limits to a series of dynamic ecological changes. To all this, one more detail needs to be added. At the center of my time in the North Coast was a collaboration I undertook with the Gitga'at First Nation. For three years, I worked with their chapter of the Guardian Watchmen, collecting digital records of their traditional territory. At the beginning of each month, when the weather allowed, we would motorboat out from Hartley Bay. I would sit in the back with our two teenaged field technicians, Steven and Ethan Dundas, sharing earbuds and listening to A$AP Rocky as we bounced across waves past rain-blurred shorelines. By documenting all the life in the coastal archipelago—as recorded from the sounds of birds, whales, waves, wolves—we worked to generate an aural record of environmental change for the nation's decision-makers. We called it the Gitga'at Ambient Baseline.[71]

The baseline is not something I discuss at length in this book. In the protocol agreement I signed with the nation, it was requested that I not use it as an opportunity for ethnographic study. It is an encountered limit to this account, a silence that nevertheless speaks to the systems of accountability and friendship in which this research was embedded. These systems are central to the way the North Coast is "experienced, known, evaluated, and critically interrogated" in *A Resonant Ecology*.[72] They hold me to the region in other ways than I recount here, involving other sounds and other stories, and many friendships. But for all this, the baseline raises some questions about which I can speak: What sorts of local histories should researchers seek to bring to community engagements? To what extent can contemporary experiments in sound making and listening enable desired political collectivities? With its architectures of metadata and code, the baseline continues the hegemonic accountings of place and ecology prescribed by the development we sought to challenge. Can new potentials nevertheless be found from within capitalism's contexts of rapid technological change, as Walter Benjamin famously argued for his own time?[73] Perhaps baselines invite opportunities to listen against the closures their methods might insinuate. Perhaps this baseline will be a basis for future perceptions of the North Coast, community-led and utopian in ambition. Sorting out the prospects for such possibilities is a task I consider in the remainder of this book.

Chapter Outlines

With one exception, the book's five case studies proceed in a chronological sequence, following the course of my fieldwork engagements. In their presentation, they also mean to build into a sense of the interconnected ways in which sound informs the dynamics of North Coast development: how the aestheticizing tendencies of environmentalism (chapter 1) help to provision a technoscience that attached itself to whales as bearers of environmental risk (chapter 2); how a noise abatement campaign delimited by neoliberal rule and settler nostalgia (chapter 3) operates against the prefigurative noise celebrated in an Indigenous popular music (chapter 4). With the final empirical chapter (chapter 5), I present something of synthesis—a decidedly false one—in the form of an eco-governance project (Smart Oceans) that seeks to assemble the cast of North Coast perspectives noted into a hegemonic vision of sustainable marine development. Let me now address each chapter in more detail.

Chapter 1 considers coproductions of whale song at Cetacea Lab. Through an ethnography of the lab, I consider how institutions shape ways of listening—including historical forms with embedded presences up and down the North Coast. Thinking with the aesthetic regime Theodor Adorno terms "late style" affords me the opportunity to consider how whale song—with its histories of New Age practice, musicking, and the domination of nature—extends possibilities for institutionalizing the senses, while expressing divergent histories of human-cetacean relations.[74] Deeply felt investments in whales on the North Coast form at a time when degrading animal habitats around the world are being marked by new kinds of investment/concern. Yet they are being delivered through a conservation spectacle whose very mediated qualities deepen its loss—premised as they are on the delimitation of the nature they seek to liberate. But whale music's ontology is also bound up with mediations that transcend human historical forms.[75] At Cetacea Lab, the mediations of whale music reveal complex and multifaceted listenings: attentive to lateness and crisis ecology, to institutional dynamics and cultures of technoscience, but also to the ceaseless and spiraling inventions of nature itself.

Chapter 2 moves from listening human subjects to the new ecological problems confronting acoustically sensitive whales. Worldwide, the vulnerability of whales to ocean noise has forced shipping capitals into acknowledging ocean noise as an acoustic-biological risk and an economic hazard, especially in regulatory hotspots. Rather than address the problem, ancillary markets in technoscience are rendering acoustical impairment as an assistive pretext, that is, impairment as a frontier for capitalist-led innovation.[76] As I show through readings of disability studies and political ecology, noisy listening has become a productive power offered freely by whales, with technoscience organizing "work as a multispecies process."[77] Pursuing this idea across cartographic projections of North Coast marine space, chapter 2 links ocean noise to industry science, to state regulatory inertia, and, later, to the hidden yet enabling work of shipping labor. Today, a liberal regulatory politics making ocean noise's impacts more visible to some is contributing to the occlusion of the spatially distributed work of many others. This calls for a new science, coupled with a new labor politics, of ocean noise.

Chapter 3 moves landside and into the small ex-fishing community of Dodge Cove. Beginning in 2007, the conversion of Prince Rupert's Fairview Terminal into an intermodal container port has subjected residents to the abstract sounds and accelerating rhythms of economic globalization. The chapter's point of departure is the sensory logics that emerged as residents sought to contest an encroaching spatial envelope of industrial noise. Building on Henri Lefebvre's concept of "state space" and Mishuana Goeman's "settler grammars of place," chapter 3 argues that residents' listening practices reveal rich ecological place attachments alongside increasingly tenuous claims to land.[78] In Dodge Cove, residents feel hemmed in by the future and the past: by the prefigurative activities of port expansion, and the cherished local rhythms port activities could dissipate. Instead of a "revolt of the living against abstraction," local turns to listening will tighten a possessive sense of place, an imagined bulwark from capitalist anesthetics that is in fact constitutive of development logics. The result is a story of doubled sonic forgetting: state space, and the coloniality upon which it operates.

Chapter 4 moves into the visionary spaces of music. It relates a history of Gyibaaw, a local musical project that cultivated critical forms of Indigenous self-recognition through the idiom of black metal. In conversation with family members, friends, and Gyibaaw's founding musicians (Spencer Greening and Jeremy Pahl), I look at how teenage passions, "fugitive aesthetics," extractivist development pressures, and living connections to territory would mark a short but eventful career.[79] Drawing from critical theories of Indigenous art, I consider how Gyibaaw's music was mediated by globally ascendant sound cultures and locally honed opacities—a combination that would prove tragically amenable to white ethno-nationalist rearticulation.[80] This speaks to music's incomparable polysemantic powers, as well as the challenges of ethical listening. To consider Gyibaaw through sonic materialism is to discern the rumblings of the conjunctural, while striving to recognize that concurrent temporalities are echoing out in other ways, for other audiences.

The final chapter (chapter 5) considers multipurpose acoustic ocean observing as an expression of new enclosures on the North Coast. Smart Ocean Systems is a state-supported effort to synthesize diverse regional interests into an inclusive network of monitoring and environmental risk preparedness. Since 2014, Ocean Networks Canada (ONC) has established a range of regional community observatories in key development areas of the North Coast. Drawing on theories of enclosure from Álvaro Sevilla-Buitrago and Mark Andrejevic, I consider how the sonic objects, practices, and discourses of ONC's Smart Oceans presage the emergence of new socialities on the North Coast and, with them, a normalization of digital-cum-maritime enclosure.[81] Central to this analysis are digital sound and the consequential mediations digital sound enables. Smart Oceans is erecting the constituency deemed necessary for Canada's sustainable marine development. As it promises new capacities in the face of unknowable future shocks, it facilitates the loss of once-integrated socialities and the sensory knowledges they helped cultivate.

What is to be done about all of this? The conclusion considers prospects for liberatory environmental politics on the North Coast. After reviewing the book's major arguments and findings, it considers the formation of new research networks in Gitga'at Territory and beyond. Alongside the hopeful possibilities of the Whale Sound project, I consider the rise of an Indigenous rematriation movement that seeks to return stolen songs and decolonize the ear. I close with country music, a cultural form with surprising breadth of appeal in the North Coast, a place where "country belongs in no country"; and a place, perhaps, where new alliances can be forged from new sensibilities.

1

AT CETACEA LAB WHALE SONG AND CONSERVATION'S "LATE STYLE"

At first, there is nothing but hiss. Then, a bright sonic flutter, lasting several seconds. What comes next might be described as two long gestures, interspersed with pauses. In those pauses, the echoes are clear. It is hard not to intuit a purpose: the animal composer listening back, interpreting, sounding out the surrounding space. As featured in the online article "Caamaño: The Sound of (Whale) Music," this sample of humpback whale song is short (just over a minute and a half long).[1] I know it as part of a much longer display, since I was at Cetacea Lab during the period in which it was recorded—as was Darcy Dobell, the media strategist whose text seeks to profile and celebrate the remote part of the coast where Cetacea Lab operates.

When I first arrived at Cetacea Lab, its mission appeared to me both firm and fixed. It was a sentinel for listening, a beacon of conservation-in-action. The tiny station was regularly being profiled by journalists, environmentalists, and policymakers up and down the coast. Cetacea Lab's rise had been swift. It was only a few years earlier, in the early 2000s, that Janie Wray and Hermann Meuter began encountering humpback whale songs for the first time.[2] "This was remarkable because humpback whales were thought to have been extirpated from the North Coast waters for decades," Janie told me.[3] In 2004, they cataloged thirty-five individual humpbacks. By 2006, the number had doubled. A total of 543 individual humpback whales were recorded in 2018, more than enough to convince many conservationists that the waters surrounding the station should be designated Critical Whale Habitat under Canadian law.[4] For Hermann and Janie, this would be a timely development, given the untimely news of regional development proposals and the transformative changes development was posing to their research.

In this chapter, I consider the different kinds of mediation humpback whale song arranges at Cetacea Lab, including the receptions Dobell celebrates in her article. More than illustrations of a passionate environmentalism forged in sound, these mediations tell a story of conservation's changing institutional field. The changes I would witness at Cetacea Lab over my four years there are a story of how, as structures of norms and rules, institutions inform particularized ways of listening. The visiting interns whom I met at Cetacea Lab listened to support its conservation science. Some also listened to gain a validating personal experience, one affixed to an economic imperative: to "inspire caring, which translates into commitment, which leads to action, which is assumed to be productive."[5] As I explore below, this latter moment spoke to the new relations of conservation spectacle that were forming along the North Coast in those early years of development politics. At Cetacea Lab, it would intercalate with still other valorizations of singing humpbacks. With his concept of *late style*, Theodor Adorno signaled an interest in art's capacity to articulate lateness; the lateness or moribundity of existing social orders.[6] Late style, I will wager, captures something about the aesthetic guiding Hermann and Janie's cetacean listening practices. It historicizes their project and connects it to an emotionally charged conservationism unique to this part of the world.[7] It denotes an interest in nature's prefigurative loss, as expressed in humpback song. It helps to explain an enduring depth of investment in the listening act; with late styles' imbued sense of, to draw from Edward Said's gloss, "being at the end, fully conscious, full of memory, and also very (even preternaturally) aware of the present."[8]

To consider Cetacea Lab's varied listenings (scientific, experiential, aesthetic) is to consider how a conservation-focused project internalizes what is in fact a powerfully unsteady combination of mediating interests, each bound to distinctive logics. "That which enables anything to make sense," notes John Mowitt, "reaches well into the institutional field."[9] The institutional focus is overlooked in geographies of sound, which tend to focus on the body, the territory, or even the planetary as preferred sites. Here, it is useful to momentarily return to the article at the top and ask: What kind of institutional project does this propose? Dobell's piece was written for the Hakai Institute, the conservationist arm of the Tula Foundation, a BC-based private nonprofit rooted in the marriage of tech entrepreneurism and cutting-edge conservation science (and which funds several entities that fund Cetacea Lab). What her article suggests is a conservation built not on oppositions, such as might restrict whales from human interference, but connections. This is a conservation fit for the neoliberal moment, one that transposes promises of alterity with instrumental institutional objectives—discouraging ruminative aesthetics in favor of spectacle-driven listening modes. Whale song will carry this move, with all the institutional pressures it solicits. But more than a singular imposition of what is, ultimately, a development process, my time at Cetacea Lab would also affirm whale song as something else: a distributed object, without original or facsimile—but flourishing variations that speak to the creativities of humpback whales.

"Just Listen": Legacies and Lineages in BC Whale Conservation

Katy Payne, Paul Spong, Helena Symonds, Janie Wray. The experimental fringes of whale conservation have long been a space where experimental listeners have pushed against dominant strains of scientific objectivism. Theirs is a project which demands that officialized conducts of observation cede space to empathetic and even intuitive approaches, and in some cases, different domestic and working arrangements too. Janie Wray is the lead researcher at Cetacea Lab. Since cofounding the station with Hermann Meuter in 2001, she has dedicated her life to whale song, often listening alone here for weeks on end, miles from any human community. For this, the whole southern tip of Gil Island (also called Whale Point) has been acoustically reengineered: the lab's front room is a hub for its regional network of hydrophones, headphones, and mixers. They broadcast sound across the tiny outdoor speakers that have been threaded along forested pathways and into the back building. Cetacea Lab interns quickly grow accustomed to kicking out of their tents at 3 a.m. to respond to a whale call. Where technologies and human senses do not suffice, other actors are employed. Neekus, one of Cetacea Lab's two canine residents in the early years of my visits (he passed away in 2016), had been trained to bark upon olfactory detection of whale scat, so as to alert humans to the possibility of an encounter. Cetacea Lab's is an almost delirious investment in listening. Addressing an arriving batch of interns on a bright summer day in 2014, Janie elaborated on its ethos: "When you sit and put those headphones on and start to listen to [whales] communicating with each other, close your eyes. If there is something strong near your nose, move it away. Envision that you are in the water. So, there is all this dark space, and then all of a sudden there are these sounds moving through. Just listen. That, on some level, is what it is like for a whale."

Just listen. Since the 1970s, humpback whale songs have played a formative role in the evolution of human listening practices—musical, scientific, ecological, and otherwise. Whales aver Steven Connor's point that animals "play an indispensable, though often ignored, part in our care of the senses."[10] Just as we mediate the world through the senses, humpback whales "mediate our senses to us."[11] In this way, we develop new listening capabilities, and with animals, new institutionalized cultures of listening too. Humpback whale songs are perhaps the paramount instance of conservation's aesthetic mediation of sound. There are countless books (many of which I've read) exploring this relationship. They discuss how songs deliver powerful conversion experiences and lead young men and women to devote entire lives to the work of protecting whales.[12] There are books, and there are also living devotees like Janie Wray and Hermann Meuter, who have constructed Cetacea Lab as a testament to the validating powers of cetacean sound.

This appeal rests on a legacy of liberal environmentalist aesthetics. But it also rests on that legacy's congeniality with the epistemological orientations of cetology (whale science), a field with long-held convictions about underwater sound. In 1966, two of the discipline's founding theorists, William Schevill and William Watkins, observed, "We are pretty well restricted to acoustics for underwater measurements and observations at any distance greater than a few meters, for of our available sensory paths only sound passes well through water."[13] Years later, this has remained the case. "It comes down to what you can measure and record," Andrew Trites, director of marine mammalogy at the University of British Columbia, told me during an

FIGURE 1.1. Listening at Cetacea Lab, 2012. Photo by the author.

interview in his Vancouver office. "Acoustics is probably the simplest and most consistent thing that can be done."[14] Acoustics lies at the basis of Cetacea Lab's conservationist ethos: to generate data from listening, in support of its effort to convince the Canadian government to designate the region as Critical Whale Habitat, but also to change people—thus saving both from industrial development.[15]

But this simplicity of recorded outputs does not equate to a simplicity of inputs. Humpback whale song is a wondrously complicated scientific proposition.[16] For cetologists Luke Rendell and Hal Whitehead, it is the most compelling case of nonhuman culture in the entire animal kingdom. "The progressive changes that humpbacks make . . . to their songs are so rapid and extensive it is hard to think of what else could account for them," they note.[17] Vocal variations in humpback songs have been used as evidence of cetacean memory, innovativeness, and sophisticated spatial analysis.[18] While there are many other remarkable animal acoustic cultures, beginning in the 1970s, song appraisals in particular would find a broad uptake through the cross-fertilizations of popular music and popular environmentalism.[19] For white middle-class communities in particular, the mediations of musical form facilitated new determinations of an intrinsic species worth, paradoxically, alongside a new interest in whales within mass culture. Singing humpback whales entered exclusive domains of human civilizational worth—intelligence, creativity, economic value—to a considerable degree through the sounds they made, which were circulated far and wide, and celebrated in various ways.[20]

One of the listeners enabling these developments was Katy Payne. Until recently, Payne was largely known as a supporting figure for her ex-husband Roger Payne. Roger Payne's celebrated 1970 study—the first formative proposition of humpback whale song defined in terms of acoustically discernable and hierarchically organized themes and phases—features her as a sidekick.[21] But his discovery rested to a considerable degree on Katy Payne's listening. Katy Payne was not a trained scientist. But she was a trained classical musician. To make sense of the sounds her husband sought to study, Katy Payne had turned to the sonata form—with its three-part structure of development, exposition, recapitulation. Janie Wray deeply admired Katy Payne, noting how her practice combined experimental ethics with "years of patient and repeated listening." Speaking to me from her home in Ithaca, New York, Katy Payne vividly recalled the sense of reflexive engagement the whales conveyed: "Well, you definitely know that the whale is listening to its own song and has decisions which are population wide about how each portion of the song is changing with time. . . . The song itself is an ever-changing pattern of, you'd almost say, 'decisions,' and all the singing whales that we recorded from a certain place and time were adhering to these decisions."[22]

Reviewing this history helps us grasp the appeal that continues to circulate at Cetacea Lab. For Katy Payne and for researchers who have followed in her wake, whale song carries models of interspecies spatial relating that challenge modernist dualisms of subject and object, body and environment. Song is a relentless stream of cetacean innovation that manifests as a continual branching, producing a proliferation of distinctive song variants. For Gregory Bateson, shiftless forms of humpback whale song were an expression of "extended mind." Here, song is not ersatz resolution, but late style in its complexity and movement.[23]

Cetacea Lab is a research station. It is also a kind of Earth spaceship. Remote and self-powered, it is an expression of the environmentalism that began to populate remote landscapes of North America in the 1970s, combining crisis ecology with New Age communitarianism. A third institutional model is also on display here, the recording studio—a project defined by its acculturated investments in documenting and experiencing sound. As

Louise Meintjes reminds us, recording studios can be forceful sites for projecting cultural ambition.[24] As a recording studio, Cetacea Lab captures a range of interests in the aesthetics of environmental music—institutional experiments with field recording, stage/setting, fidelity, and the prospect of nonhuman musicking. In this sense, it captures one of Adorno's criteria for late style—engagement with historicity of musical forms. It is through the recording studio mediation that whale song delivers its evident capacity to enact musically meaningful environmentalism. When David Rothenberg writes that "we would never have been inspired to try to save the whale without being touched by its song," he is speaking to this sonic community.[25] He is speaking to Janie's eco-musicalization of humpback whales, a practice all Cetacea Lab interns are taught to cultivate. This is a playful environmentalism, but as a history of human-cetacean relations, it is also mediated by something less exultant. In Janie's practice, it is imbued with a sense of lateness—a lateness that spans a gulf between individualized feelings and the outward appearance of fading natures; between "embodied sensations and feelings as psychic or cognitive experiences," to quote affect theorist Ann Cvetkovich.[26] We find it throughout the cetological history Janie and Hermann relate: in Scott McVay's 1966 lament about the waning of "the great whales," in Roger Payne's 1995 career retrospective, and in the more recent expostulations of Tom Mustill, who compares cetacean encounters to "read[ing] by the light of a library as it burns."[27] Whale song evokes a history steeped in understandings of human destruction. As a recording studio and listening practice, Cetacea Lab mediates several historical feelings, some uplifting, some not.

During rainy stretches, when singing whales were absent, I would sometimes wonder what had driven Hermann and Janie to come all the way to Gil Island. Here they were, without scientific accreditations or money, in a landscape where weather can keep you landlocked for days. But in addition to its eco-musical ideals, whale song has long nurtured a lifestyle promise that resonates here. In BC, there has long been a cottage literature celebrating whale research as a kind of sentinel practice: young men and women, white and able bodied, venturing into the wilderness to discover whales and themselves. Books like Farley Mowat's *A Whale for the Killing* (1972), Rex Weyler's *Song of the Whale* (1986), Erich Hoyt's *Orca: The Whale Called Killer* (1984), or Bruce Obee's *Guardians of the Whales* (1992) tell tales of hydrophonically mediated whale encounters on misty mornings, along windswept beaches, and near abandoned village sites. Lingering animal voices appear as remnants of once-massive prewhaling populations. Sometimes, they inspire longer-term partnerships—sentinel stations like OrcaLab, or the Centre for Whale Research, or Cetacea Lab itself. In concretizing such visions, individuals get to live out public fantasies of nature immersion. Lateness figures a kind of asceticism, an environmentalism of self-imposed exile. As explored in the Dobell article, such efforts can also become objects through which nature experiences can be broadcast to the world.

This is significant as regards local development politics. Here as well, we find the substitutive logics that have long marked ideologies of nature in British Columbia, with singing whales both symbolizing and ensuring the silencing of colonized First Nations (who tend to be heard in the whale narratives only as voices from the past). When I met Hermann and Janie in 2011, it was apparent that they were doing much to acknowledge the territory in which they operated. They spoke gratefully of their Gitga'at hosts and how the late Chief Johnny Clifton's support "was the reason we can still be here," as Hermann put it.[28] But Hermann and Janie were also concerned with how the station's new attention was affecting local relations. There were so many visitor requests, not from Hartley Bay, but from the myriad other communities with whom they had lately become connected. What was becoming increasingly evident to Hermann and Janie was that many of these communities did not recognize the weight of their territorial responsibilities.[29]

Cetacea Lab is modeled on OrcaLab, a whale research station established in 1978, three hundred miles south of Gil Island. Hermann and Janie met there as interns in 1992. Its architect was Paul Spong, a brain scientist and flautist who is the hero of Weyler's 1986 book *Song of the Whale*. On Hansen Island, OrcaLab championed the noninvasive listening-based approach that various research institutions up and down the coast (and, to a certain extent, in New Zealand and Australia) have since adapted and elaborated. OrcaLab is an argument for the geographical specificity of whale musical mediation, the promise of live encounter through local encounters with musical sound.[30] It is also a space marked by an aesthetics of loss, a feeling to be both suffered and cherished, and whose logics would be repeated in Cetacea Lab's built form and cultural ambitions.

Spong, who is widely credited with having convinced Greenpeace to launch their Save the Whales campaigns, agreed to host me for a few days in the fall of 2013.[31] We met on a desolate gray morning at the Port Hardy harbor, and I got inside his tiny skiff. Much like its junior station up north, OrcaLab peers out from the forested mist like a hallucination: an oceanside perch built of curved beams and stained glass, nestled between thick lashes of cedar and hemlock. The Earth spaceship ideal is strong here. More than an aesthetic choice, the mandalas and solar panels are reminders of the apocalyptic earth narratives that first drove Spong and many others to the fringes of the BC coast. "My beliefs about animal captivity in the 1960s were not made through the training I received," Spong said in his quiet voice, as we sat drinking tea on the front porch. "They were made through the whale I listened to." We talked about the pods of orca in the waters around us. Like his former interns Hermann and Janie, any mention of whaling or of present risks of ocean noise filled Spong with a visible unease. Emotionalism was at the basis of his ethically charged turn against mainstream science. In Spong's telling, conservation needs to extend not only from awareness of ecological crisis but also from the human spiritual crisis born of diminished attunement to the world.[32] The failing might be ameliorated, if never fully rectified, through musical listening. As far back as 1972, Spong was describing whale music as a system for "the communication of emotional data" with "significant potential in the realm of interspecies communication."[33]

Spong's critiques of a cetology divested of listening's empathetic capacities have made him a "difficult beast for some" (his words). While beloved by environmentalists and journalists, OrcaLab has had rocky relationships with the whale research community and the Canadian government, to say nothing of industry. Perhaps an avowedly noninvasive research approach, with only intermittent interest in publication, makes this opposition to the mainstream unavoidable, I suggested, given the different institutional pressures it involved. Spong demurred. He wondered aloud if younger generations of regional inhabitants were not becoming less interested in whales, in spending real time with them. "We are losing whale habitats, and are we not losing them too," he asked me, "the actual effort to become dedicated partners?" While it is hard not to find some of this persuasive, it is also hard to avoid context. There is much in Spong's eco-philosophy that seeks to challenge Western dualities. But at OrcaLab, whale song aligns with a "bourgeois ideology of nature"—an ideology whose power extends from its capacity to simultaneously appraise nature as external and universal, and as an experience that is beautifully consumable.[34] Here, nature's ideological contradictions are pursued in the privileged belief that sound "brings us closer to everything alive."[35]

Adorno elaborated his late style from the same postwar Los Angeles context in which Spong had trained as a scientist (he completed his PhD at UCLA in 1966). For both figures, LA was far from the center of European high culture, if also riddled with the excesses of capitalist consumerism. As David Jenemann argues, a sense of dissonance, fragmentation, and dislocation—born from these circumstances—marks much of Adorno's West Coast writings (much as it did for his fellow Frankfurt School expats).[36] In the years following Adorno's death (in 1969), and right around the time of the release of *Songs of the Humpback Whale* (1970), a new generation of white middle-class men and women sought exit from its moral-spiritual decay—and the growing calamities evident in human relationships to nature. Spong was very much a part of this history. But sitting with him at Orca Lab, forty years later, it was evident that Spong's listening featured something more than a late-style commitment to endings. At OrcaLab, late style might well index a consciously produced sense of exile—both geographical and historical in nature. That this sense could be paradoxically assayed in a remote and purifying nature (and one built from acts of Indigenous erasure) speaks to contradictions of bourgeois modernity the late-style aesthetic affirms. But the whales one listens to at OrcaLab are not exiled figures, and Spong knows this. Rather, the whales are home, or, within one of the many homes whales make in their ocean movements. At OrcaLab as at Cetacea Lab, whale song's presentation as sonorous artifact of an extant but fading nature, as something yet intimately tied to the violence of western culture, confronts a human historical limit: the realization that here other senses take place, and take place in ways humans may never know.

The Hydrophonic Hiss

"Listening *to* whales," a Scottish intern named James told me candidly one day, "is more listening *for* whales. They usually aren't there." What is there is the sound of the hydrophone, hissing, gurgling away, bringing flows of underwater space into our awareness. The hydrophone is always there, bubbling like a coffeemaker at one moment, or the quiver of a record player at another. Adorno was fascinated by the hissing record player. In this sound, he discerned what he called the truth of mediation: the fact that modernity impels second-order technological experiences that both shape and substantiate experiences of the natural world. At Cetacea Lab, the hydrophone is the central mediator of human acoustical life. It is the basis for Cetacea Lab's Passive Acoustic Monitoring (PAM) research, intercepting soundwaves and converting (transducing) them into voltage for airborne detection.[37] For Hermann and Janie, as for Spong, using hydrophones upholds a normative commitment to whales—"passive" as opposed to "active"; "monitoring" as opposed to "intervening."[38] To engage whales through hydrophones means a "commitment not just to whale research but whales as well," as Hermann once put it. In 2001, he and Janie had just one hydrophone, a homemade unit loaned from a friend down south. They would canoe to the middle of Taylor Bight and drop it in the water "just to have an ear" (Janie). When I first arrived at Cetacea Lab, hydrophone-based listening involved a multi-channel recording rig. You adjusted knobs (pans) on a large mixing board to foreground different channels corresponding to the four hydrophones installed around in the region (Taylor Bight, Squally Channel, Whale Channel, and Caamaño Sound). In 2013, Hermann and Janie acquired two Ocean Sonics icListen hydrophones, priced at about $10,000 each. They came through the philanthropic giving that had swelled in the wake of Enbridge.[39] In 2014, four more listening locations were added, giving the station expanded coverage in Ursula Channel, Verney Passage, and Otter Pass, as well as new possibilities for aggregating data and triangulating calls.[40] Cetacea Lab's older analog hydrophones had not been calibrated, meaning they offered incommensurable magnitudes when compared against one other.[41] This had been fine for the lab's interest in simple abundance statistics. But it was inadequate for the kinds of collaboration envisioned by Hermann and Janie's new partners—Ocean Networks Canada (ONC), the World Wildlife Fund (WWF), Save Our Seas, and the Department of Fisheries and Oceans. In the North Coast, the digital hydrophone is a story of the expanded concerns of whale conservation. It is also a story of the expanded power of data analytics, digital labor, and sonic capital. It is, to a considerable extent, a story of development politics.[42]

Cetacea Lab's move to a condition in which whale song was to be captured by automated sensors, compressed, and transmitted to satellites and stored in the cloud did not happen all at once. Rather, it evolved in a piecemeal fashion. The process would be marked by new funding windows, personnel availabilities, and the enduring physical challenges of the locations where they wanted to install new units. But Cetacea Lab was far from alone in expanding its tools: In 2015, I conducted an informal survey of five other nongovernmental whale research stations along the BC coast—SIMRES, Salmon Coast, OrcaLab, Pacific Wild, and Strawberry Isle. All confirmed similar investments in digital hydrophones at roughly the same time. At Cetacea Lab, the most remote of the stations, now centrally figured in the crosshairs of a development spectacle, these investments displayed hallmarks of the technological modernization that was reaching up and down the coast in new ways.

During my third summer at Whale Point, a disappearing sound seemed to imprint these changes with a strange poignancy. Clipping is what happens when an audio signal is amplified past the maximum allowed limit. It is undesired from a scientific perspective: generally, scientists want good (i.e., measurable) frequency response, which clipping compromises though its delivered effect of overdrive. During my first two seasons at Cetacea Lab, Hermann and Janie actively solicited clipping. Before going to bed, they would boost the gain on their outdoor speakers so that loud sounds would max out and produce audible distortions. They did this to ensure that opportunities for whale detection were not missed during sleep. One result was the experience several interns would gamely recall as the "3 a.m. wake-up call." Moments before the signal broke, clipping was informing us that a humpback (or a killer whale) was singing nearby and ready for monitoring on headphones. With hydrophone digitalization, the *schzzzzz* sound was no longer necessary. I never heard it during my fourth season at the lab because there was now software that could automatically record signals that were amplified to a certain level, bypassing the need to test the physical components of the speaker on us.

With the loss of clipping, a small but meaningful marker of activity and intentionality at Cetacea Lab was gone, an event that had mediated and expressed a desire to encounter whale song in a particular way. Clipping hardly qualifies as an object worthy of nostalgia. Many cetologists appreciate the affordances digitalization allows—everything from the increasing flexibility of study design to the augmented size of the audio datasets they can store. But researchers openly wonder about the unaccounted costs of research digitalization. Paul Spong told me that when less time is needed for actual listening, there is "less opportunity for finding real connections"—of the sort that informed his ethical orientation. Not unlike Adorno, Spong promoted a philosophical commitment to the particular; a *lingering* with it. His long-time collaborator and life partner, Helena Symonds, told me that an automated research approach would obfuscate the goal of trying to live more intimately with nature: "I think one of the most satisfying experiences in life is to be in the lab, and you are there, and you are trying to understand what's going on, and there's all these groups out there. And it goes on and on. And then it's over and you realize that you've been there for this whole several hours of life, movement, and activity."

In a 2021 paper, cetologists Eduardo Mercado and Christina Perazio argued that information theoretic analyses of humpback whale songs (which recommend digital hydrophones, owing to the copious data they require) obscure "how singers are changing song features over time by discounting acoustic details."[43] They can hide otherwise discernible interests in rhythm as an organizing template in humpback song organization. At Cetacea Lab, these debates offer windows onto the changing relations of science that have transformed observation-based conservation around the world. The 2010s were the decade when machine listening approaches entered mainstream institutional usage.[44] It might one day be remembered as an incipient period of human-machinic exchange, when different logics worked to find points of agreement and cross-purpose, much as they appeared to be doing at Cetacea Lab.

The effects of technology were equally if not more transformative in another sphere of daily listening activity. Throughout my time there, representations of Cetacea Lab's headphoned listener were a commonplace theme in antishipping environmentalism. A spate of recent documentaries—including *Oil in Eden* (2010), *SpOIL* (2011), *Tipping Barrels* (2012), *Groundswell* (2012), *The Great Bear Rainforest* (2017), and *The Whale and Raven* (2019)—sought to harness the region's natures for advocacy purposes. In their routinized portraits of Cetacea Lab, they sought an "affective aura"—one that could incite viewers to feel and care in the structured ways Cetacea Lab's featured listeners evidently did.[45] These efforts evoke Dobell's blending of musical environmentalism and scientific musicalism. With song, it is not simply place that needs to be conserved, but the fleeting artifact of nature that conservation alone can protect. Each becomes a site of viewer/consumer attachment. Like the song sample in her article, the film's samples do this work efficiently. Song scenes are never more than five seconds, enough time to guarantee the Althusserian obviousness that is proof of their ideological character, but never so long as to detract from other sites of investment.[46] Several days before my first visit, an MTV film crew had been on site, filming *Pipeline Wars*. "What was it like to listen on camera, with a film crew pointing at your ear as you listened?" I asked one of the interns. She didn't really know, she said. But then she added, "It made things feel a bit faster than if they weren't there." The intern's sense of harried uncertainty contrasts with the communicative gestural listening modeled by the MTV reporter, made visible in the film.[47] Upon encountering whale song at the station, she first registers shock, then turns to the camera and exclaims, "That sounds like a David Guetta soundtrack!"

The increasing presence of what Jim Igoe calls "spectacular conservation" would have integrative effects on the listening practices Cetacea Lab pursued.[48] As Guy Debord famously observed, spectacle does not pursue relations of dialogue.[49] It does not valorize listening of the type Michel Chion once called "semantic mode"—where formal curiosities about songs are allowed to exercise themselves. Nor does this conservation satisfy his threshold for "casual listening"—the simple marking of sound's appearance, such as to mark animal presence.[50] What *Pipeline Wars* models instead is a kind of instant recognition: gratified and possessive.[51] In this, it upholds an imaginary of neoliberal conservation: that a region and its natures can be saved through the consolidating structures of networked individuals inspired to demand change. This imaginary solicits a listening without durative temporality, one that delimits the questions that might otherwise unfold: Is this song something the whale takes pleasure in singing? Is its structure changing because of other dynamics, like the encroaching presence of warmer waters? In accepting neoliberal conservation, Hermann and Janie would accept practices that posed jeopardy to the slow commitments they sought to cultivate. *Pipeline Wars*, like so many products of its anxious moment, only needs to communicate that nature experiences are *happening*. It projects a clarity that holds little value for the embeddedness of the experience, its micro-gestures of self-identification, relation, environment.[52]

The promotions of whale song by the NGOs and large media organizations are not purposeful attempts to de-valorize other listening practices. But their efforts to ensure a particular received effect must nevertheless be understood in terms of an expanded commodity logic. Digital communities have the form of commodities; they can be bundled and sold to advertisers. "Dramatic performances" of conservation activity can happen; and they can happen as consumable, calculable, listening experiences that will motivate new interests in whale conservation.[53] In the 2010s, the effected link between conservation and advocacy-entertainment blossomed into carefully curated discourse: environmental media communication. As I witnessed firsthand, Cetacea Lab's funders and partners would routinely ask Hermann and Janie to furnish their promotional efforts with a communicative intensity. "If you experience this place," Janie intones in a 2016 web video posted by the WWF, surrounded by ocean mist and lush vegetation, "you will be driven to protect this place." In 2010, Cetacea Lab was one of the first whale research stations to livestream its whale recordings to listening publics. By 2016, this iteration of "wired wilderness" was commonplace.[54] Online visitors from around the world can tune in to flows of whale migration at places like Whale Point and participate in the "making of global connections."[55] The song gets scattered globally—sucked through hydrophones channeling the ocean waveguide and flung into networks made of far-off reception points. I sometimes scanned the content posted on Cetacea Lab's Facebook page, the song clips eliciting hundreds of enthusiastic comments, likes, reposts, and links.[56] As Lab-based interns work to ensure the feeds are kept up, they become increasingly committed to their maintenance of these networks and not the ones around them.

Like Janie, Cetacea Lab's more public spokesperson, Hermann Meuter, is deeply charismatic. But his demeanor is more unpredictable, and he can be surprisingly candid in extended conversation. Hermann told me he despaired of the social media labors Cetacea Lab was being asked to do. A former semipro soccer player from Stuttgart, Germany, Hermann had journeyed to OrcaLab on a whim. There, he met and fell in love with Janie. In the early 2000s, their romantic relationship ended, but they continued to collaborate on Cetacea Lab. But Hermann had become despairing of the ways the station was being branded. He told me about the random emails he received from ecotourists looking to promote the station by way of a free afternoon visit. He yearned for the period when he had been able to immerse himself in the sounds. "The most abundant period for us was between 2006 and 2010," he told me one day. "The songs in Whale Channel used to last for hours into the night. We just sat there and listened." A detail surprised me. During the years I had been there, it was Caamaño Sound, not Whale Channel, that served as a focus for the lab's efforts. This is what the Dobell article suggested. Like Lelu Island, Kitlope, Gwaii Haanas, and Khutzymateen before it, Caamaño Sound was being branded as the next Eden on the North Coast that needed saving. "Perhaps Whale Channel was just not spectacular enough?" I said, half-joking. Hermann laughed. "In this part of the world, that would have been saying something." I proposed a different topic. Had he and Janie ever tried to identify individual humpback singers, I asked, much like they could do with photo ID of their flukes? "No," Hermann said at last. "That would have been very difficult to do, and we would have had to get out there . . . and that would not have been appropriate. This whole thing, for us, was about listening to them, not disturbing them in that process."

Hermann's concerns over intrusion, something he saw as relevant to my question, reminded me of Spong's own refusals. During the documentary filming periods, or when large troops of visitors were there, whales, too, might be feeling uneasy, Hermann later explained. The human accoutrements of spectacle—the bags of recorders and sensors, the relentless movements, the hours spent waiting to capture the perfect moment—all of these were inconsiderate extractions of the spaces in which whales acoustically made place. Intruding in the composition was something Hermann was unwilling to do; even talking about it made him upset. As the conservation NGOs pursued greater connection and access, he was moving in the other direction, increasingly concerned with the strange discontinuities of whale song. I came to understand this as a position that had cost him considerable support in the conservation community, costs he was still working out, years later.

Conservation is failing, environmentalist and former Nature Conservancy head Peter Kareiva has written, because humans are "increasingly disconnected from nature, and as a result less likely to value nature."[57] The sentiment has only grown more commonplace in the years since Kareiva wrote those words (2008). By and large, desires to promote nature experience have only deepened NGO investments in the communicative media that promise ever greater nature integration—pixelated, high-fidelity, virtual natures, if not physical ones. While expressly concerned with biophysical spaces, the agreements that established the Great Bear Rainforest in fact support this logic.[58] They support a whale conservation that pursues its mission alongside various kinds of human presence, not absence. They support acts of saving nature alongside nature exposure activities—like ecotourism and commercial fishing—and thus, economic growth and even development.[59] What, I wondered, does all of this mean for the contemplative listenings pursued by Paul Spong or Janie or Hermann; the listenings that strive to be "fully conscious, full of memory"?

"Something feels different this summer," Janie said one evening. We were sitting around a campfire by the beach: four interns, myself, and Janie. It was August 2015, my second to last season as a Cetacea Lab visitor. I had noticed something that week myself, a perturbation in the collective mood.[60] Was it anxiety over another looming visit from one of the millionaire prospective donors staying at nearby lodge? Was it to do with Hermann, upstairs in the lab listening to '80s German pop on his headphones? It was also true that we hadn't heard whales for over a week. Years later, I would learn that the time of my visits (2012–15) had been marked by some impactful ecological shifts. Bioecologies up and down the North Coast had encountered a range of disturbances, including the arrival of a debris field from the Tohoku tsunami (2011), a large El Niño Southern Oscillation event, and a Pacific Decadal Oscillation switch that contributed to record-breaking surface temperatures the year before. One of the resultant effects was a sudden decline in the humpback whale population's calving rate. It is possible that Janie had been attuned to some of this change. For Janie, humpback whales' life stories are actively inscribed in regionally composed whale songs. The idea is akin to Bakhtin's "utterance," with each song containing echoes of past voices, past songs.[61] A talented former intern named Eric Keen later published work that lent support to Janie's convictions about humpback place-making. The aesthetic qualities of the North Coast's underwater fjords are a "motivating factor" for the visiting whales, Keen wrote.[62] They come here to sing, and they come here to communicate the losses and the changes they have experienced, season after season, year after year.

More can be said about these historically minded insights. Less than 100 kilometers from Cetacea Lab, tucked inside a sandy cove at Kunghit Island, are the remains of Rose Harbour. Established in 1910, the former whaling station was once responsible for some of the most intensive harvesting up and down the West Coast.[63] Humpback whales, the first species to be hunted in British Columbia, would have been dragged up a long wood haul-up slip. Their blubber removed by flensing crews, they would then be winched onto a carcass slip, where bones, meat, and viscera would be extracted and processed. The bodies came in the hundreds. Janie told me that the humpback whales who visit the North Coast today know this history. In ways she cannot substantiate scientifically, they have returned to the North Coast "because they feel safe here again," she explained. It is an intuitive claim, but there is no reason to assume it is false, given what cetologists have learned about the historicity of whale song's production. History—human and animal—is made up of rhyming structures. It is possible that the insights we gather from future whale songs will reveal communities whose challenges bear an eerie semblance to today's. Rose Harbour is now an ecotourist guesthouse where kayakers can explore tidal meadows and the remains of Haida totem poles. But who is to say it can't return to its violent past?

Conclusion

This chapter has considered humpback whale song as a distributed object, with changing interrelations between its component mediations. At Cetacea Lab, these different moments combine into a story of development politics, a story about how one multispecies community has navigated some of the disruptive changes development enacts. A whale music understood to be cetacean in origin is also produced by human desire, human technology, and human loss, I have shown. At Cetacea Lab, humpback whale song thus reveals something about how institutions function, and how Cetacea Lab in particular would function as an institution in transition: between political economies and attentional modes, technologies and ecologies, calls and responses. I have offered late style as one artifact of this transition; a historically minded listening being displaced by other forms. But this is not a lamentable thing necessarily. There are other, more radical ways of listening to whales today—engaged with new questions of nature and humanity, with whales "as a form of life that has much to teach us about . . . vulnerability, collaboration, and adaptation."[64]

Much has changed at Whale Point. What started as a couple with a tent and a hydrophone in 2001 transformed by the middle of the 2010s into an internationally recognized project, with funders, partners, social media followers, and a battery of marine acoustical tools. Compelling studies about area abundance and habitat use would be published. New partnerships would take shape with local and international collaborators. But tensions between Cetacea Lab's two founders would deepen across this span too. Eventually, the development dynamics would provoke an insurmountable rift. In 2017, a messy institutional breakup occurred.[65] Janie left Whale Point to start her own research institution, BC Whales, on nearby Fin Island. Hermann continues to work out of Whale Point, but the name Cetacea Lab is gone. He has turned away from the conservation groups and the research ambitions of his former organization. By the outset of the COVID-19 pandemic in 2020, Hermann's Pacific Whale Society had become an important space for marine educational engagement among Gitga'at youth, who could visit from nearby Hartley Bay. By all accounts, the last few years have been a challenging time for both Hermann and Janie. Nevertheless, they continue to share song recordings as part of their lifelong commitments to the whales of the region.

Most visitors to Cetacea Lab have some understanding that ocean warming, acidification, plasticization, and diverse sources of harmful anthropogenic noise are all steadily making their way into the waters around Whale Point. In invisible ways, slow and perhaps deadly violence is seeping into the lives of humpback whales, possibly affecting how and what they sing, possibly inducing changes that will become measurable only later. All of this is an occasion to listen, and to grieve. Late style, Robert Spencer reminds us, is less about an individual composer than the social order being revealed in their work—one whose presentation requires a great many acts of mediation.[66] But there is a twist at the end of this story. It returns us to the whales who first brought Cetacea Lab to worldwide attention. Despite myriad ecological concerns and unaccounted costs of extant industrial change (including more barge and ferry traffic), humpback whale populations have been doing well on the North Coast. In 2012, the numbers were considered low. Observable population changes began slowly but grew into a strong evidence base by 2018—when Cetacea Lab, in its last official year, recorded 543 individual humpback whales. As overall conditions of regional ecological health track downward and people anxiously await news on shipping proposals, whale song carries a different story—a countermovement, perhaps—to the declensionist arc of a "late conservation." When I last visited Cetacea Lab, it was for a quick drop-in on an afternoon in July 2018. Approaching the station from a Guardian survey vessel, I suddenly worried that we'd miscommunicated. I was relieved when Hermann finally appeared on the outcropping to wave us ashore. The humpbacks of Whale Point were doing better than they had in decades. The humans, less so.

2

VALUE IN INJURY THE WORK OF SCIENCE IN OCEAN NOISE REGULATION

When the propeller blade of a large ship rotates through the water, the propeller's suction and an interacting pressure force create tiny vapor bubbles or cavities. The result is ocean noise, the sound of thousands of bubbles collapsing, a signature of the vessels now responsible for moving more than 90 percent of the world's goods.[1] Ocean noise catalogs shipping's global political economy: its trade volumes, more than quadrupling from 1970 to 2019 (from 2.6 to 11.1 billion tons); its vessel numbers, increasing by more than 80 percent during the same period; the carrying capacity of its average vessel size, more than doubling.[2] Driven by capital's relentless need for growth, ocean noise is helping to realize a significant transformation in the sea: from heterogeneous environments once defined by the periodic calls of the great whales, into marine monocultures defined by chronic broadband noise.[3] In some parts of the northeastern Pacific, acoustic tracking technologies suggest that ocean noise has increased 12 decibels (dB) since the late 1960s alone.[4] Ocean noise would receive little economic consideration for decades. But insofar as shipping is the source of most of the world's ocean noise (upward of 75 percent, by most estimates), it is the audible expression of an epoch-making abstraction—capitalism's ongoing reconstruction of diverse ocean spaces into "lines drawn by boxes shipped across the globe."[5]

At Cetacea Lab, Janie Wray encouraged me to imagine ocean noise from up close: deep inside the bodies and minds of whales. Interns were invited into this idea whenever the sound of motors of nearby fishing boats rang out from the speakers. Affixed to a cinematic world of antidevelopment conservation, the humming motor sounds brought intimations of cetacean suffering and "preverberations" of violence to come.[6] Listening, we contemplated the idea that whale song would vanish; that whales would "go away and not come back to these fjords," in the words of one scientist.[7] For humpback whales, Janie explained one afternoon in 2013, the significance of ocean noise goes far beyond the increases in decibel levels that the NGOs and government were finally beginning to consider. Rather, the real concern lay with the qualitative changes to the formal structure of the entire region's sound profile: "Ocean noise means new *kinds* of sound," she noted. "It means sounds entirely different than the ones whales are used to or have evolved to understand."

Janie was not alone in this assessment. By 2012, the biological hazards cetologists were reporting in connection to ocean noise had become extensive: everything from heightened whale stress (measured in cortisol levels) to area-avoidance behavior (as whales fled noisy areas), to permanent hearing loss, to reproductive failure. The idiom through which state regulators would interpret these findings would be one of "risk"—a knowledge practice through which various institutional interests can be brought together. Regulators were by 2012 just beginning to consider the risks that whales might, in turn, pose to North Coast shipping. Massive payloads of chronic ocean noise were set to become a dominant feature in the still quiet, undershipped waters of the region. Janie drew me a dire picture: individual calves cut off from their kin by a pulsing industrial din; deadening clamors that would induce invisible injuries, errant behaviors, even strandings—sound from which there was no escape. It was one of the reasons Cetacea Lab had partnered with the WWF to construct an ambient noise baseline from the hydrophones they had recently set up. The recordings could give them new insights on how whales were coping in what scientists were increasingly calling "the Anthropocene ocean."[8]

How do whale species, with their distinctive *Umwelten*, actually experience ocean noise? What kinds of relationships do they pursue when entire sensate environments turn hostile? Sarah Besky and Alex Blanchette conceive scientific labor as an "attunement to other species' rhythms."[9] The principled work of Janie and Hermann suggests an example of this theory, with situated knowledges carefully derived from listening. But Besky and Blanchette's theory is not an inherently progressive one. As I show in this chapter, the ontological mysteries of nonhuman listening have also become of interest to capital, and a capitalist technoscience finding new forms of value generation in the prospect of acoustically injured animals. Through ocean noise, capital is assembling new kinds of multispecies work, mediated by new kinds of observed injury, biophysical change, and risk.

What kind of work is this? It is work, I argue here, hinged to efforts to grasp "other species' rhythms" inside a well-shipped North Coast. It is also work that seeks to contour map, risk assess, and policy recommend ocean noise without challenging its enabling logics. I draw on a combination of animal geography, disability studies, and media studies to develop this argument. I consider how ocean noise regulation, and its role in the various kinds of work making up an ocean noise assemblage, reveals foundational problems in efforts to apprehend and mitigate the impacts of ocean noise. Of particular interest to me is the relationship between underwater acoustics and whales, and the perspectives of different individuals working at this nexus. This focus routes my analysis away from transnational flows of shipping and toward "the transnational ligaments of technoscience."[10] This yields important insights, but as I reflect at the end, it also results in some missed opportunities as regards the broader development dynamic on the North Coast.

Science, in this telling, provides a key vantage for making sense of the discouraging regulatory gridlock that has gathered around ocean noise worldwide since the early 2000s. This is to a certain degree because ocean noise is an exemplary "boundary object" in Tom Gieryn's sense.[11] It has long been grasped as a problem requiring multiscalar and cross-agency management solutions, given its transboundary movements and transdisciplinary effects.[12] But for all the institutional efforts to recognize it as a global biological hazard, ocean noise remains a story of trenchant local asymmetries—a hazard that slips between the cracks of different government agencies, corporate actors, and knowledge communities. On the North Coast, ocean noise outlines a neoliberal world of science in which commercialization and development politicking tie regulators in "rhetorical and practical knots."[13] It illuminates a situation where regulators must manage a crazy quilt of industrial project areas—with their distinctive timelines and biophysical impacts. More than any figure, it will be whales, and killer whales in particular, who explain the politics of ocean noise and consolidate its assemblage of actors on the North Coast. Theorizing the role of killer whales as both risk and opportunity in this projective economy is thus a useful way to begin the account.

Technoscience and Assistive Pretext

For Mara Mills, the history of disability science in the North America has pivoted to a significant degree on a "rehabilitation model" of injury: Could an individual experiencing disability still work (and if so, how well)?[14] Mills's concern is with human subjects and the "humanity" accorded to differently racialized and gendered bodies. But her ideas are helpful for grasping the plight of injured whales in an era marked by the "biopolitics of the more than human."[15] Elizabeth Johnson has forcefully shown how capitalist technoscience works through practices that valorize nonhuman life in terms of "what it can communicate to *humans* about the vulnerability of life in the material world."[16] The noise-sensitive whale is of a piece with this story. Through whales, technoscience considers a boundary line of acute interest to both state regulators and shipping capitals: What separates acoustic conditions conducive to marine life from those deemed destructive of it? On the North Coast, the observed behaviors and abilities of whales, along with their measurable responses to sound, will affirm the political nature of ocean noise regulation and serve to highlight this regulation's reliance on risk to measure and mitigate disability, or injury (the term more commonly used). In turn, all of this will help to explain why one of Mills's significant conceptual offerings, "assistive pretext," should appear at the center of my critique too.

"Assistive pretext" is about the "resourcing of disability by technoscience."[17] Key to the concept is Mills's insight that institutions working within advanced capitalist societies routinely project disability as a "precursor and pretense"—that is, as a justification for research funding, or lucrative markets involving nondisabled peoples.[18] In the case of ocean noise, assistive pretext will point to the cadres of acoustical consultants, software engineers, and data analysts that occupy positions in the "ocean noise assemblage."[19]

These actors are less visible than the whales and the scientists that focus the issue, but they mediate their engagements in key ways. As regulation stalls and institutional commitments to sustainable marine development advance with little material effect, the biologically expansive problem of ocean noise becomes translated—via intermediary material-discursive practices of sonically focused data analytics—into new frontiers for capital accumulation.

As a tool for ocean noise critique, assistive pretext leads us away from the frameworks preferred by the mainstream regulatory communities. Such assessments affirm global shipping as chiefly responsible for ocean noise production but offer little with which to interrogate it. Rather, shipping is reified as ubiquitous but impossible to dissect, responsive only to carrot-and-stick approaches that might reduce ocean noise at zero cost, or perhaps even benefit.[20] Regulators have little more than voluntary commitments to show for these efforts, decades later. The momentum always seems to break on the shoals of the international regulatory body—the cadre of organizations (like the International Maritime Organization [IMO] or the Center for Biological Diversity) where shipping regulation has proven largely ineffective, and for reasons best understood elsewhere.[21] Against this vantage, assistive pretext invites us to consider shipping's enabling supports in new ways; including those purportedly neutral consultants and analysts who are profiting handsomely from poor regulation in development contexts. An attention to these actors can help us grasp the contours of a related configuration: a technoscientific capitalism more invested in sensing the conditions of ecological demise than in responding to such conditions on a structural level.[22] This does not mean that direct critiques of the shipping economy are not also needed, but rather, that we need new ways of approaching their interconnections. In what follows, I outline an assemblage of ocean noise actors and link it to the dynamic of assistive pretext on the North Coast. I begin first with a history of expanding technoscientific relations between whales and ocean noise, before turning to the important innovations of acoustical data science and the space of conjunctural shipping politics.

The Informatic Acoustic Ocean: Signal Traffic and Big Data

In the late 1990s, whales emerged as the unofficial mascot of ocean noise risk. This is a story I will unpack in several ways, as it is necessary to cast a wide net to grasp the different institutions and actors who have influenced this outcome. Ocean noise risk, moreover, can be used to explain several consequential outcomes, including the unruly combinations of gridlocked regulation and advancing technoscience I have already mentioned. If we are to recognize its significance on the North Coast, there is a crucial additional factor to consider: development. As Joost Van Loon observes, risks are actualized in anticipation.[23] Here, ocean noise's status as a quantifiable hazard, or risk, has been consolidated in relation to a particular kind of futurity—industrial activity in a series of to-be-highly-shipped channels and waterways. Ocean noise is not only a risk on the North Coast. But it has achieved special salience in this context because of its ability to articulate a complex of new policy requirements, ecological inventories, and technological supports—in short, development's coming boom.[24]

The postulate of ocean noise did not leap from a blank page, unnoticed one day and graspable as risk the next. Rather, its story begins in the annals of Cold War oceanography and naval interests in a contrasting potentiality: an ocean "not-noise," an acoustically uncluttered soundscape wherein the sound signatures of moving vessels could be observed from great distances. For Cold War oceanographers, ocean noise mattered first because underwater sound was a privileged medium for vessel detection and tracking, including enemy submarines. Ocean noise interfered with ocean signal. This idea rested on emerging understandings of marine space as "a natural acoustic signal rich in frequency, diversity, temporal variability, and directionality."[25] One of the enduring legacies of Cold War oceanography, in other words, was its conception of marine space as an audible formation, constitutive of different spatial-temporal envelopes that mediate different kinds of signaling.[26] This conception developed as the result of myriad discreet acts of scientific labor, including some that took place not far from the region where I began my own research, decades later.

In the fall of 2014, I traveled to Victoria, BC, to meet Ross Chapman, a soft-spoken former naval officer with the Canadian Department of National Defence. From the late 1960s until the 1990s, Chapman was part of Canadian oceanographic efforts to measure ocean noise in the North Pacific. His job involved listening to thousands of audio recordings and classifying the different noises he heard—rain, wind, earthquakes, shipping. The work was slow and recursive: "We tried hydrophones at different depths," Chapman told me. "Measuring the sound made by breaking waves . . . listening to the sound of rain and other phenomena for hours, trying to explain it all in terms of the physical principles."[27] Long before a smart ocean was ever conceived, military actors had rendered the North Coast into an instrumented field of sonic detection, with retrievable sonobuoys bobbing atop the waves of its outer reaches. In this way, naval scientists "acquired massive reams of data about the Cross-Pacific. Transits of hundreds of ships of all types." What happened to all those recordings? I asked Chapman. "They are somewhere in the bowels of the Department of National Defence offices," he replied, laughing. "There are probably thousands of dusty cassette tapes with ocean noise."

Necessary to ocean noise risk was the conceptualization first established here: the ocean as a profuse but finite signal space, a medium that could be rationalized for purposes of managing signals.[28] Since the 1970s, oceanography has parameterized underwater sound transits in terms of amplitudes, frequencies, and phase fluctuations. In so doing, it has provided a framework for conceiving of underwater sound in rationalistic terms. This would be significant for several reasons. One reason has to do with cetology (on which, more in a moment). Another has to do with shipping, and the fact that naval oceanography did little to upset the imaginary that shipping capital and various other capitals have privileged across most of the twentieth century. This is the idea of the ocean as a "void into which unwanted things could simply disappear."[29] To a signals-obsessed oceanography, noise was an unwanted thing, not a biological hazard. As early as the 1940s, Chapman's predecessors were classifying shipping as the ocean's dominant anthropogenic sound source.[30] They heard its low-frequency hum as part of a jumble of signals that also included many living things: groaning fish, snapping shrimp, singing whales.[31] These attunements mattered little. Or, as Chapman summarized, "We know about the whales *and* the noise. But what we knew then barely registered as worthy of concern, at least from an environmental standpoint."

Cetology would break this compact. In the 1970s, cetology embraced acoustics, mostly famously in the form of humpback whale song. This led to new speculations about acoustical "interference" between competing ocean users, or what the US and Canadian navies discussed as "spectrum crowding."[32] These insights emerged because of a novel scientific attunement, including from such notable figures as Katy Payne and Roger Payne (see chapter 1), who had friends in the US Navy toting strange underwater sound recordings. For the Paynes, underwater sound was about more than signal and noise. It was a basis for "the daily work and complex networks through which animals reproduce themselves and their communities."[33] Through cetology, but progressively expanding into other marine biology, underwater sound, the sensory cue that travels farthest through the ocean, became reappraised as a vector for registering changes in the ocean's capacities to support marine life. Some of the earliest studies into ocean noise risk involved sightings of gray whales, who seemed to avoid acoustically intensive petroleum exploration activity.[34] Then, in the 1990s, a series of beaked whale strandings were conclusively linked to nearby high intensity military sonar. This galvanized new scientific concern over the acoustic impacts of anthropogenic marine activity and supported the turn to other sounds. In the early 2000s, a half century after planet-straddling propeller cavitation was first observed by oceanographers, international conservation bodies claimed sufficient confidence to add the chronic low-frequency sounds of shipping to the welter of anthropogenic acoustic threats.[35]

This backstory helps establish some of the key details that continue to define the ocean noise assemblage. Central among them is a significant and enduring reconceptualization of whales: from once-silent swimmers to dynamic webs of acoustic signals, built by and responding to various kinds of environmental information, including noise.[36] Scientific engagement with the ocean noise–whale relationship would prove to be remarkably significant for the building of new knowledge coalitions and the establishing of new questions at the nexus of acoustics and biological impact. Whales' capacities to perform various kinds of metabolic, ecological, and affective work in relation to different anthropogenic noises—from sonar to seismic air-gun blasting to propeller cavitations—would, in turn, recommend their central place within the regulatory discussions that had emerged in response to the science in the early 2000s. By this point, ocean noise had ceased to exist as simply a biological or ecological issue. Rather, it had become part of an expertly administered "world ocean," marked by new annual conferences, ISO measurement standards, lobbying efforts, popular appeals, and policy commitments.[37]

A set of institutional configurations had thus transformed the ocean noise assemblage and brought the issue into new prominence by the time I began my research. These realities were on full display at a workshop I attended at the Vancouver Aquarium in February 2013. "Ocean Noise in Canada's Pacific" was organized by the WWF. I was fortunate to have been given an invitation based on some mutual contacts in Prince Rupert, specifically Janie Wray. Entering the conference room, I encountered a who's-who of ocean noise science in Canada: prominent cetologists like Lindy Weilgart, Christine Erbe, and Lance Barrett-Lennard; Erich Hoyt, a key architect behind the emerging discourse of marine protected areas; Jack Lawson of the Department of Fisheries and Oceans (DFO); and Michael Jasny of the Natural Resources Defense Council (NRDC), whose policy reports in the late 1990s were the first summaries of growing scientific work on shipping-related ocean noise. Sitting almost directly across from me in the large boardroom was Janie Wray, the sole invitee who wasn't affiliated with a large organization.

The WWF conference was a "trading zone."[38] It was a space for epistemic enhancement from the various cognitive communities seeking to better understand ocean noise. Over the following two days, I sat and listened as different delegates considered the pertinent issues, including the growing presence of noise across coasts; the global state of cetacean injury modeling; and the prospective noise increases in the North Coast from the Enbridge Northern Gateway, at the time a point of acute environmental concern. For my purposes, two details stood out. The first involved the sound map we were shown—one of the first of its kind. It displayed recent-recorded levels of ocean noise in the North Coast using Automatic Identification System data of ship movements and diving marine space into 250 × 250 meter grids. Shipping vessels were grouped into different length classes and sound source spectra—that is, the amount of vibration at each individual frequency. Values are assigned to the classes based on insights drawn from literature reviews and published reporting. A statistico-taxonomic grid, the map conveyed orderly management schema—a method, perhaps, for measuring the spatial-acoustical changes ocean noise would introduce. In this way, it marked a passage point between Cold War oceanographic engagements with ocean noise (e.g., Chapman) and new regulatory interests lodged in data analytics: a shift in using statistics on underwater sound as descriptive quantification into one fundamentally about predicting outcomes. More than a summary of decades of recording efforts, the sound map was a signal of an emerging regulatory reality: ocean noise to be managed as a problem of data modeling. All the situation now required, to quote marine scientist Rianna Burnham, "was to put the whale in the map."

The second detail pertained to shipping. WWF's ocean noise conference confirmed the arrival of an era in which shipping could no longer be ignored—on the North Coast, but in other contexts too. There were no shipping delegates in the room that day, but the industry was present on the sound map the research team led by Christine Erbe had produced. It was not a definable set of actors and interests but a series of thick red lines, routes that designated "persistent acoustic repetitions in the ocean," as ocean noise policy advocate Michael Stocker later put it to me. Critical cartographers have long argued that naturalization is one of the greatest power effects maps can produce. Here, in red lines, was the intransigence of the shipping economy, part of the broader story of shipping's naturalization across marine space. At the end of the second day of the conference, I raised this idea to Janie. She responded that it reminded her of Hermann's experience at the Marine Planning Partnership for the North Pacific Coast (MaPP) earlier that

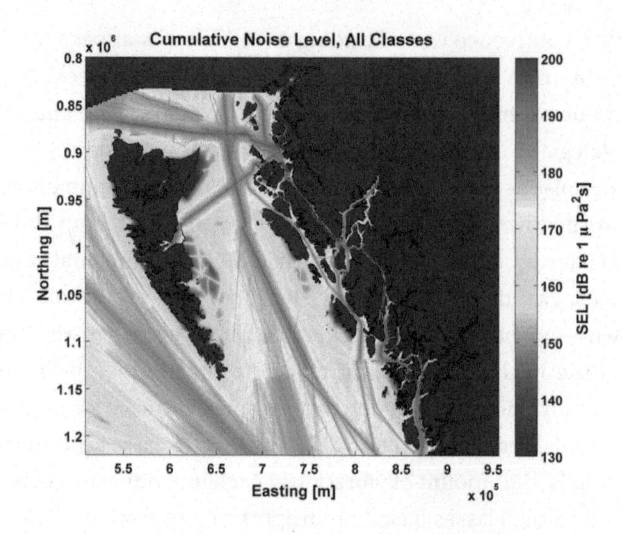

FIGURE 2.1. Cumulative sound exposure level from vessel traffic, January–December 2010. From Erbe, Duncan, and Koessler, "Modelling Noise Exposure Statistics from Current and Projected Shipping Activity in Northern British Columbia."

year. The MaPP meetings had been organized by the BC provincial government to address ocean noise along with a host of other ecological concerns. At MaPP, Hermann learned that any possibility of rerouting shipping away from whale-abundant Squally Channel was "totally off the table, despite possible alternative routes." Red lines were also there, in effect. They would also materialize in the routes that future LNG proponents would, again, largely determine for themselves.

These two features—a technoscience, rooted in data analytics and prediction, and a global shipping industry, seemingly impervious to regulation—would contribute to the ocean noise assemblage and make it amenable to assistive pretext. They would interact productively, inhibiting the regulation at the source (e.g., the ship's cavitating propeller) while redefining the work required to determine impact at the receiver (e.g., whales). And the particular concept that linked them was risk, a term that had been routinely invoked at the Vancouver conference.[39] Ocean noise qua risk was a suturing between the concerns of biological scientists and the cost-benefit models of capitalists. It legitimated a technoscience with capabilities to itemize and pinpoint degrees and spaces—of risk. Michael Stocker put it to me this way: "Industry-funded scientists assume that there are acceptable limits for an exposure . . . a point of *risk* . . . and proceed as if their task is to determine a pathway to a goal line—one which may or may not actually exist biologically." Growing institutional interests in risk maps, noise budgets, and quantifiable trade-offs between vessel speed and habitat exposure would impart the rapidly consolidating institutional understanding: any mitigation must be able to adjust for the acoustic space required by shipping. Shipping must be accommodated, not contested.[40] And for this to work, whales need to accommodate, in turn.

"Putting the Whale on the Map"

Recent years have seen a wealth of critical geographical efforts to make visible the ways animals remain indispensable to the mutations of capitalism.[41] For Maan Barua, animals are "workers in the shadows of capitalism."[42] The reference to visibility is instructive. Not only do nonvisual animal capabilities matter in the case of ocean noise, but the work whales do in this context is hard to visualize—hard to measure, premised on further improvements in modeling, and largely buried amid the considerable reportage ocean noise would begin to receive along the BC coast in the early 2010s.

For scientists and regulators alike, putting the whale on the map—giving whales new visibility as objects of risk and concern—meant drawing new attention to the question of listening. Cetologists concur that killer whales receive most of their sound through their jaw, which contains a fatty tissue called acoustical lipids. From here, sound is channeled to the auditory bulla underlying the jaw, which vibrates and in turn conveys sound into the cochlea, the inner ear, and eventually the brain. If this model seems straightforward enough, it is also the point at which scientific understandings of cetacean listening meet their limit. The problem has been long recognized with regard to ocean noise. "The problem in determining the biological significance of marine mammal responses," a National Research Council report observed in 2005, "is that often we do not know them when we see them."[43] As a 2007 paper by Cara Horowitz and Michael Jasny elaborates, "How does one know when a powerful noise source has compromised a whale's ability to detect predators, or separated it from its calf, when that whale is underwater 10 miles away? What does it mean for a humpback whale to change its song patterns, or for a sperm whale to alter the way it dives? Getting to the bottom of these questions," the authors conclude, "will take years—even decades."[44]

For killer whales, a species that long held commanding interests from marine conservation along the BC coast, ocean noise risk can manifest in synergistic, behavioral, and physical effects. It can be community-wide, and in other contexts, shaped by individual capacity. For these reasons, cetologists identify a range of oftentimes contrasting killer whale responses to ocean noise, including "area-avoidance," heightened levels of stress, and unexpected tolerance.[45] Added to these are questions of changing predator-prey interactions and changing calling patterns to conspecifics—themselves difficult to measure and with undetermined long-term effect.[46] Reputed cetologists like Lindy Weilgart, with whom I discussed the issue at the Vancouver conference and in follow-up phone calls, stressed the uncertainties: "We are at the beginning of this issue, not the end." Weilgart went on to tell me about different kinds of ecosystem stresses and the need to be vigilant in the face of ambiguous findings. Much to her chagrin, the epistemological uncertainties of the noise-whale relation have not been met with precautionary regulatory responses. Rather, they have become purposed to a technoscience governed by risk and the search for conditions under which industrial activity can proceed without sufficient overlap into whales' bioacoustical niches.

There have been several forces encouraging this approach. One is the marine-based industrial capitals themselves—shipping, oil and gas, seabed mining—who have found new reason to become involved in ocean noise science and regulation since the early 2000s. The logic, as the NRDC's Michael Jasny explained to me, is simple: "The industry may be in situations where they are barred from going into certain areas because regulators won't let them. Or regulators might privilege one company over another because of a higher ability to reduce potential impacts." As a result, these capitals have become major players in a science that fuses data analytics and oceanographical acoustics with the minimal guardrails of state-led environmental assessment. "Risk assessment is driving a lot of the [research and development] work that we are seeing," Jasny continued, "and intensifying efforts to find quieter alternatives to area restriction." In the North Coast, where risk might include the legal penalties proponents might accrue by harming whales, ocean noise invites a virtuous cycle of risk-technological improvement.[47] Given the federal government's staunch support of shipping, local regulators must be predisposed to "innovative solutions." In broadcasting an "urgent need for real-time information on the occurrence of marine mammals," regulation renders noise into lucrative opportunities for technoscience.[48]

Like other Western nations with large maritime economies, Canada has undertaken aggressive efforts to support its oceans technology sector in recent decades. In so doing, it has implicated ocean noise management with broader efforts to catalyze dynamism in a multibillion-dollar Ocean Supercluster.[49] This initiative, and others like it, is also meant to bolster Canada's regulatory bona fides—most notably an *Oceans Protection Plan* (of 2016) that extends a suite of promises for sustainable marine development. By the mid-2010s, a range of funding opportunities and strategic partnerships were newly available for small and midsize firms interested in developing new software and tools for the quantification of ocean noise. Here as elsewhere, socio-technical capacities to assess risk will be shaped by different institutional priorities, economies, and projective sites and spaces of assessment. But after more than a decade since its commitments to the mitigation of ocean noise were first tabled, it is evident that in Canada, regulatory activities confront the limits imposed by the sustainable development horizon.[50] Rather than a singular injunction, what I would find on the North Coast was a set of actions and exchanges whose overall effect is best summarized as assistive pretext.

A summary scan of the emerging acoustical consultants sector is helpful here. By the time I began tracking the sector in 2013–14, notable firms working on ocean noise regulation included JASCO, Greenridge, Seiche, Intelligent Ocean, Bae Systems, and Colemar. Smaller firms included Ocean Sonics, Sonsetc, RTSys, Sea Mammal Research Unit (SMRU), and Wildlife Acoustics. Whereas smaller firms typically specialize in one or two activities, such as measurement and modeling, the larger ones are often both manufacturers and service providers. Many have extensive histories with large militaries (which became an early funding source for ocean noise research in the 1990s in the context of litigation on navy sonar).[51] Besides building and selling instruments (including the sonobuoys, hydrophones, and gliders I encountered on boat rides with the Gitga'at Guardians), an emergent area of work here is behavioral response studies (BRS), in which paid ship-based observers report behaviors of nearby whale populations. Firms now offer expensive courses to qualify individuals as marine mammal observers (MMO) as well as operators of PAM systems with other certifications.[52] A single field season of BRS can cost an industry client hundreds of thousands of dollars.[53] Passive acoustic monitoring, often done in tandem with BRS, is another increasingly lucrative service area. Industry interest in PAM has spurred the growth of firms like Ocean Science Consulting Limited, SMRU, and Smultea Sciences. The international acoustical consultant Seiche, which dominates the PAM services market, offers a range of PAM services to its clients—including towed systems, remotely operated monitoring, and various training courses.[54] Each of these markets is predicated on advancing capacities to parameterize and quantify whale responses in relation to real and hypothetical noise exposures.

Scientists rarely listen to the sounds that whales experience. But in efforts like these, cetacean listening becomes a productive power that whales freely offer, and one for which they receive little in return (e.g., as mitigation).

Perhaps the most lucrative growth area in this economy has been in acoustical software and artificial intelligence (AI). A good early example of ocean noise modeling is Population Consequences of Acoustic Disturbance (PCAD), which estimates the energetic implications on whales of masking—that is, noise-induced communication interference.[55] Building on state-led initiatives in the United States—including CetSound and CetMap—PCAD visualizations follow whales through dynamic and fuzzy models of marine space, which can likewise track the movements of ships. A PCAD software package released by the SMRU in 2014 was hailed as a "significant step forward in the ability to assess risk to marine mammal populations," specifically by being able to spatialize their changing risk geographies.[56] Since 2014 or so, acoustical consulting firms have become more interested in exploiting advances in signal processing related to machine learning and AI and now regularly harvest sonic data from public ocean datasets.[57] One problem these firms commonly encounter concerns data hyperabundance. The proliferation of new hydrophones in places like the North Coast has resulted in reams of acoustic data that need sifting through. As a result, AI-based approaches to acoustic data processing have become another site of growth activity. Here, efforts have focused on two tools in particular: detectors and classifiers. Both are used to put whales on the map. They allow consultants to address questions such as When and where did we detect whales? What kinds of noise levels did these detections correspond to? Detectors help regulators establish patterns of whale-ship correspondence that facilitate conditions of acceptable risk. Classifiers go one step further in being able to categorize the specific sounds a whale makes in response to reported noise levels (whether a moan is higher or lower pitched, for example, and what this might reveal about acoustical conditions).

Given all of this, it is worthwhile asking how the ocean noise assemblage has changed institutionally and which institutions have benefited. Established in Victoria, BC, in 1981 by Joseph A. Scrimger—a scientist whose previous career experience included the Defence Research Establishment Pacific—the acoustical firm JASCO now operates subsidiaries in the United States, United Kingdom, Germany, and Australia, and employs over one hundred people. As an acoustical services company, JASCO is an illustrative case: its growth has almost entirely occurred within the decade when ocean noise became an international regulatory concern. More than a service provider,

JASCO is an ocean noise intermediator, realizing sonic capital through its ability to articulate various mediations—encompassing whales, state actors, industry tools, and flows of underwater sound—to new economic interests. JASCO's AMAR hydrophones, mentioned in my study of Cetacea Lab (chapter 1), would become recognized as the "Cadillac" of hydrophones during my time on the North Coast, owing to their wide frequency ranges and sensitivities. JASCO also offers more than twenty whale-specific detectors that can be programmed with its AMARS. While nonproprietary detectors exist—such as SMRU's PAMGuard—JASCO's are routinely cited as a gold standard and commonly featured in large industrial assessments accordingly. Clients from government, industry, and NGO worlds pay subscription fees to use them and then separate fees for data processing, which JASCO oversees through its "humans-in-the-loop" validators. These actors, additions to an expanding ocean noise assemblage, listen to ocean recordings in near-real time to report on missed whale detections (data that JASCO can subsequently use to improve its detectors). Validators can be serviced to provide daily, sometimes hourly, updates for JASCO clients. According to one regulator I spoke with, a single validator can cost a proponent upward of $50,000 in fees on an annual basis. There is evidence to suggest that an expanding market for ocean noise services has transformed competitive dynamics in the acoustical services sector in a number of ways, some quantifiable (such as the increasing number of scientific publications devoted to ocean noise product testing and verification), others not.[58] "If you said to JASCO in 2009, can you do an [acoustic] environmental assessment," one DFO scientist candidly explained, "they might have said, 'Sure! We have this thing we are trialing. Can we deploy it?'" The scientist continued, "At that point, they were doing R&D. But now, it wouldn't be a question that you would have an acoustic assessment. You will. And you will pay a lot more in every step."

I interviewed a pair of scientists at JASCO in their Victoria offices in 2014, finding them receptive and welcoming. When I reached out again in 2018 and 2019, I was unable to secure a reply. My claim is not that firms like JASCO are malicious or seek conspiratorial alliances with the development actors to whom they sell things. Rather, it is that they are rational actors pursuing new market opportunities. They operate as neutral providers of risky data through a logic of assistive pretext. Their detectors and classifiers privilege assessment, and not mitigation, as their site of technoscientific innovation. But their continued ability to operate is nevertheless premised on a particular regulatory culture and the "free gifts" whales offer though mediations of recording, representation, and data (indeed, acoustics firms can undertake sophisticated assessments without ever having to encounter whales or their acoustical habitats in physical space). Like the other acoustic consultant companies with which it competes, JASCO engages the ocean noise assemblage less as a set of relationships needing repair than as a material basis for the commodification of its products and services. This does not leave the whale unchanged. In the assemblage moment where expert analysis meets acoustically sensitive nature, nature becomes partitioned via algorithmic determinations, and an integrated sensory agent (i.e., whales) reappears as a composite of receptor points. Whales become statistical (and profitable) in ways that are fundamentally about predicting what they will do and assessing them in terms of risk. And all this is pointing to dismal futures for whales in an industrializing North Coast, the subject to which I now turn.

Life in an Ocean Noise Hot Spot

From my home base in Prince Rupert, attempting to track the politics of ocean noise felt like an exercise in disorientation. Here I was, in the eye of a development storm, a context central to new understandings of the environmental hazards associated with shipping—such as ballast water contamination and ship strikes. Throughout the time I lived there, acoustical consultants, state regulators, and university scientists would regularly fly in from Vancouver, Ottawa, or further afield. They would appear at industry open houses, disappear again, and six months later a regional baseline measurement was available to view online, or a risk map, or a marine mammal mitigation plan. Occasionally, I got to hear commentaries from the LNG proponents themselves. They tended to be the same sunny message: technological developments and novel technoscientific approaches were delivering what commentators were calling a "golden age" of marine data analytics.[59] The Prince Rupert Port Authority was expanding the berthing areas at its Fairview Terminal in a bid to remain Canada's third busiest port. There was a great deal of activity, coming from a range of directions. But there was also precious little in the way of coherence. I struggled to make sense of it all.

One person willing to provide me with an anchor was Mike Ambach. Ambach ran the WWF's tiny office above Pho 88, the Vietnamese restaurant on Third Avenue. He was curious about my time with Cetacea Lab and how my acoustics collaboration with the Gitga'at Nation was proceeding (at the time, the Gitga'at was a community the WWF was actively looking to work with). But most of the time we commiserated about unanswered emails to ocean noise experts. At one point in 2014, there were fourteen LNG projects exploring bids in the Prince Rupert area. Ambach had the responsibility of tracking ocean noise in relation to all of them. When I first met him, he had just authored a study suggesting that the North Coast would experience a 100 percent increase in ocean noise due to development.[60] But as he admitted to me, the trend lines were almost impossible to assess given the range of possible projects.[61] "We're having to learn underwater acoustics awfully fast," Ambach said to me during one of our downtown walks. He added, "And it's not like there's a lot of reinforcement from higher branches of government."

It was only after sustained time in Prince Rupert's development geography that I came to appreciate this lack of "reinforcement." In step with broader political mandates, state regulators in the North Coast were being instructed to treat development proponents with deference. They would not attempt to problematize their constructions of ocean noise risk through the incorporation of scientific uncertainty. Rather, they would allow the proponents to parameterize risk through technoscientific innovations that other branches of Canada's government were cultivating. An early illustration of the dynamic comes in the exposure thresholds regulators use to determine the extent of ocean noise harm. In its 2010 application to send tankers through the North Coast—an assessment undertaken by JASCO Associates—Enbridge Northern Gateway advanced a species-specific standard for estimating behavioral impacts on northern resident killer whales. Their approach built from the work of a scientific team led by Brandon Southall, whose widely cited 2007 paper makes discriminations between low-, mid-, and high-frequency cetaceans for auditory injury assessment.[62] Killer whales could thus become a functional ocean noise group based on hearing sensitivity, in this case as "midfrequency" cetaceans. For Enbridge, this meant that killer whales were sensitive to a different set of noises than low-frequency cetaceans (e.g., blue whales)—an assumption science has contested. Their standard assumed that only ocean noise 55–65 dB above the killer whales' hearing threshold would affect these whales. Accordingly, the project could reduce the extent of its risk area to an impact zone of one to two or more orders of magnitude smaller than would be calculated by other prevailing thresholds.[63]

Seeing Canada's regulators do little to question these assumptions, local conservation groups seized on the inadequacy of Enbridge's risk thresholds approach (even as few questioned the reliance on risk itself).[64] "The underwater acoustic modelling documents do not consider cumulative noise impacts nor do they factor in time, they merely present model outputs and most of this in the form of numerous one-page figures," noted one.[65] The dispute would prove to be a foretaste for what was to come. Enbridge simply sought to address the exposure of animals to a propagating sound source, independent of duration. Subsequent industry-led efforts would prove better at accommodating outside concerns, while continuing to face little headwind from the state. For example, the modeling approach in Schlesinger and colleagues' *Aurora LNG Acoustic Study* would better account for the temporality of animal vicinity: how long a killer whale is sufficiently near the source of ocean noise (e.g., the propeller). If the whale is observed to be swimming past, it is only in vicinity for a certain amount of time. If it swims closer, or if the ship moves closer, it accumulates the dose more rapidly, increasing the overall risk. Crucially, industry commitments to better parameterize risk—which built on long-running ideas of a marine soundscape to be administered in terms of rational utilization and quality enhancement—would not be met with corresponding changes in thresholds or accompanying science. Even with a strong mandate from Canada's Species at Risk Act—which since 2009 has sought to protect the acoustic habitat of the Southern Resident population[66]—Canada's regulatory agencies still lack a quantitative regimented system for ocean noise. "It's all a bunch of loose actions and efforts that are now growing year by year to limit noise," Ambach's colleague Hussein Alidina told me over the phone. "There still hasn't emerged a framework that sets a target or a threshold for noise management." At the regulatory level, project by project assessments continue, despite a raft of work documenting the dangers posed by multiple intersecting projects.[67] Between Chatham Sound (Prince Rupert) and the tip of Gil Island there exists the possibility of multiple project transits, each independently observing local noise thresholds, while producing consistent cumulative doses to whales cohabiting these waters.

To an important degree, Canada's challenges regulating ocean noise confirm the scalar challenges of trying to tackle a problem that exists at the global scale. Here, the news would also be discouraging. In 2014, the International Maritime Organization (IMO) released a set of voluntary guidelines on ocean noise mitigation designed to be applicable to the entire shipping sector. Michael Jasny, the lawyer I met at the Vancouver conference, was instrumental in summoning the conservation and NGO pressure necessary to convince the institution, which happens to be shipping's largest lobby, to finally act. When I interviewed Jasny in Vancouver in 2015, he seemed cautiously optimistic. Yes, he was aware of the *voluntary* stipulation the IMO had included, but he nevertheless saw an opportunity for cascading improvement. Given that noise can be a proxy for a propeller's inefficiency, Jasny noted, it seemed plausible that shipping capitals could pursue "dual benefits" from ocean noise mitigation: a quieter ship is not only a more efficient ship, but a less environmentally impactful one. But it has become increasingly evident that the IMO guidelines are having little effect.[68] In 2019, the WWF completed an assessment on their uptake, drawing on examples from around the world. Hussein Alidina summarized the findings to me over the phone: "It has been poor. Across the board."[69] Instead of the costly work of rebuilding ship propellers, industry proponents prefer operational measures in specific contexts. While there is some scientific merit to vessel slowdowns as well as institutional commitments to undertake them, these have so far tended to apply only in very specific spaces. The overall benefits to whales are unknown, but it is likely that any benefits will deplete rapidly when multiple, co-occurring vessels are factored in.

Some industry forecasts I came across suggested that 1,300 annual vessel transits could be a regular feature in the waterways of the North Coast by 2027.[70] The figure, which may be low, represents more than a doubling of the transits in the years I lived in Prince Rupert. Looking out at Chatham Sound from a local bar called the Ocean View, I would sometimes consider the eerie prospect unfolding below: new populations of acoustically disabled killer whales, confusedly moving this way and that, their subjective fates increasingly tethered, not unlike humans, to sensing technologies distributed in hostile environments. In the open ocean and outer continental shelf, high-frequency noise can be absorbed by seawater within about 10 kilometers, possibly before reaching a species of concern.[71] But in tight fjord-land channels like Chatham Sound, underwater sound can bounce off canyons and twist over seamounts. Killer whales may be exposed at ranges of 1–10 kilometers, even less than 100 meters in some circumstances. Scientists have no idea what will happen when these occurrences become habitual. For this reason, some scientists feel compelled to continue to listen to ocean noise, like Janie does, and like I tried to do in the summer of 2016—scanning the files cetologist Rob Williams sent me containing recordings from the heavily trafficked Haro Strait. These minute-long recordings describe a terrifying reality—entire afternoons when cetacean life is suffused with droning, jackhammering noise. Human ears cannot fully register the embodied low tones or the searing harmonics that killer whales and other species are likely discerning. Still, there was enough in Williams's recordings to jolt me from the empathetic listening proposed at Cetacea Lab ("Just listen . . .") into an entirely different space: sonic torture chambers; the "acoustemologies of detention" Suzanne Cusick explores in connection to the American war on terror.[72]

After listening for thirty seconds or so, I'd had enough. I could remove my headphones. For killer whales, the issue is they can't.

Coda: "Workers in the Shadows of Capitalism"

This chapter has explored the distressing politics of ocean noise regulation on the North Coast. Despite years of collected evidence pointing to its impacts, ocean noise essentially remains both unmitigated and unaddressed. The situation speaks to a doubled problem: ongoing efforts to mitigate impact not lessening regulatory gridlock but realizing new frontiers of accumulation through assistive pretext.[73] Following Besky and Blanchette, I suggested that scientific labor is "attunement to other species' rhythms."[74] In the case of ocean noise, attunement is also assistive pretext: an extractive practice allowing human actors to reap new profits (in the form of new tools, assessments, and acoustical services) from whales' energetic responses (listening, moving, misbehaving, etc.). In this way, the case study affirms the productive dimensions of noise. It is not simply about unwanted sounds that actors unilaterally seek to suppress, but new kinds of economic opportunity, and the making of new kinds of socio-natures as well.[75]

At the outset of the chapter, I suggested that my approach to the ocean noise assemblage has overlooked a few elements. Before closing, it is important to say more about this. Over the winters of 2013–15, I embarked on several failed noise reconnaissance missions with Marty Bowles, a retired high school teacher from Prince Rupert. Setting out from the Cow Bay Marina, we would approach the chugging container ships moving in and out of Prince Rupert's Fairview Terminal, hoping to get close enough to drop a hydrophone and listen underwater. We were never able to evade the port security detail, which always instructed us to leave. The significant fact of these missions, however, was in what came next. Sailing back up Chatham Sound, we would pass stationary container ships waiting to unload, sometimes for weeks on end until market conditions were right. If we drew close enough to one, we would hear tapping sounds coming from within. These were the ship's laborers, skeleton crews of barely five or more—their faint, percussive *tink, tink, tink* chipping rust off the ship's interior. "That's a chipping hammer, not a pneumatic chipping drill," Marty noted at one point. "It means a lot of extra effort needed." It took me a long time to register how here, hidden from sight, in shipping's own hidden abode, was another acoustic body indexing work in the ocean noise assemblage. The faint acoustical tapping inside the ship—*tink, tink, tink*—marks labor's role in resurrecting the dead labor congealed in a ship's machinery: keeping it afloat, working away, "grinding, scraping, hammering, and drilling . . . [in] an endless cycle that repeats itself every two months."[76]

In *The Right to Maim*, Jasbir Puar argues that "the biopolitical management of disability entails that the visibility and social acceptability of disability rely on and engender the obfuscation and in fact deeper proliferation of debility."[77] This statement is helpful for considering the plight of whales, and the speciesist divide Western culture upholds. But it can also be read to consider the racializing dimensions of shipping's political economy, which harmonizes, in the present case, with a marine conservation inured to the production of its racial others. A hegemonic discourse of ocean noise has made the plight of animals—whales in particular—visible in new ways. At the same time, it has contributed to the depoliticization, and obscuring, of the costs of shipping—including the physical costs borne by shipping labor. One of the reasons this chapter does not focus on shipping labor is because my encounters with ship workers were so infrequent. As in many intermodal port geographies, foreign ship stays have become progressively shorter in Prince Rupert over the last decade.[78] Rigid labor controls, combined with increased port securitization, have all but eliminated most opportunities to meet people who may spend months at a time in Chatham Sound, to disembark for only very short moments, if at all, before leaving for another port.

Still, there were moments when possibilities did present themselves—as was the case during a remarkable event in 2016, when the highly publicized bankruptcy of the Hanjin shipping line led to the stranding of the crew of the *Hanjin Scarlet* near Stephens Island, several miles out from Prince Rupert.[79] At one point, the crew was given leave to go to the Seafarers Mission in Prince Rupert, a volunteer-led organization with a storefront office on Third Avenue. A friend who worked at the port told me about the crew: they were young and mostly Filipino, and thus belonged to the community of more than 300,000 that makes the Philippines the largest supplier of global merchant labor in the world.[80] I immediately planned to go and meet them. Among other things, I wanted to know: Did they have any notion of the noise regulations connected to the ships they were on? Did ocean noise mean anything to them? And what about those tapping sounds? What was it like to experience them, week after week; to be exposed, like whales, to the force of shipping's "gargantuan automation"?[81] By the time I arrived at the Seafarers Mission, however, the seafarers were gone, back aboard the ship and setting out to their next destination.

This ethnographic failure is one I still regret. It speaks to my initial assumptions about the politics of ocean noise as well as to the analytic choices

I pursued in response. But as I explore in chapter 3, it also speaks to the ways development imaginaries can direct assumptions about what to pay attention to, and about who or what is made meaningful in sound. By 2017, more than 22,000 workers had been welcomed to the Seafarers Mission. In Prince Rupert, the situation of exploited shipping labor has remained absent in most (if not all) of the local development discourse. The absence recalls the tapping I first heard with Marty Bowles. Set against technocratic discourses of noise regulation, these sounds invite the possibility of forging new connections between the exploited workers who enable shipping's multiscale networks of trade, production, and knowledge. What's needed to affirm these connections, I now believe, is not simply a regulatory solution, but a new kind of science for ocean noise. This would be a community that goes beyond the scientific search for "opportunity-areas" that maintain the "acoustical status quo" at "little socio-economic cost."[82] It would be a community willing to speak truth to power, something Lindy Weilgart did when she questioned the navy's selective funding of ocean noise research in a 2005 paper.[83] A critical ocean noise science would be one willing to consider the expansive nature of the ocean noise assemblage for its unevenly distributed costs to human and nonhuman life.[84] It would entail a different kind of attunement, and a different kind of ethics—"an ethics that is at once experimental and that demands accountability."[85]

3

"PORT NOISE" SENSE OF PLACE AND STATE SPACE IN DODGE COVE

On April 12, 2015, I joined eight residents from Dodge Cove (pop. 52) for a hike up CBC Hill. We began in silence along the dirt road that cuts through the village, its colorful wood homes glancing through the foliage. We turned left at the forest trail and began to ascend. The freshness of a rainfall hung in the air. Our feet snapped on wet twigs, and some warblers sang. At the summit the trees thinned into tall grass, and we became engulfed in a wash of new sounds. Ahead of us, just past the narrows of Chatham Sound, stood the bright red gantry cranes of the Fairview Terminal. There were beeps from the service trucks moving about the tarmac below and pulsing signals. Every so often, a container was dropped, and a *thud* rippled across the water like a thunderclap. John spoke. "It's the sound of China." He had a knack for pithy statements. There was a Hanjin container vessel in the loading area. "That's a Korean company, John." Doug's voice. "Ah." John stroked his walrus beard. "*Shhh.*" Lou. The taskmaster of the group. But John had more to say: "Well, it might be China *soon*. At the rate they're going, who the hell's to say!" Everybody laughed, and John shot me a wink. "Hope we aren't ruining this for you there, Max!"

In this book's previous chapter, I emphasized noise in relation to nonhuman life and as a physical force. For whales, I argued, ocean noise constitutes a new form of biological hazard—while perversely extending new kinds of economic opportunity to technology firms.[1] In this chapter, I focus on how noise exerts force in the symbolic dimension. As a narrative tool, noise can be useful for coming to terms with change. Noise conflicts crystallize at moments of heightened change, politically and economically and otherwise—as Lilian Radovac and Emily Thompson have shown.[2] In such moments, noise abatement efforts can become powerful sites of collective investment and meaning-making, something I was coming to appreciate in my walks with a small community of fisherfolk and would-be noise campaigners. Since 2008, the growing spatial envelope of the Prince Rupert Port Authority (PRPA)—with upgrades to its Fairview Terminal, deemed necessary for anticipated marine traffic—has confronted Dodge Cove residents as an audible formation. Noise came to operate here as a harbinger, drawing in local sensibilities and, in turn, stabilizing and regularizing them to new rhythms.[3] I can recall standing alone one day on the Dodge Cove village dock. The air felt overstuffed with low and round mechanical sounds. If less strident than the other industrial sounds I had become accustomed to living across the harbor (where train whistles and shunting had become commonplace), there was something overwhelming about "port noise," as the villagers called it. It seemed to cancel all distance between the dock and the cranes across the water. Later, a villager offered a word for this: *superimposed*. In Dodge Cove, noise was being superimposed. For local residents, it connoted both *nearness* and *out of place-ness*. It came from somewhere larger and somewhere else but was now, inappropriately, *here*.

Dodge Cove, with its lush gardens, old boatyards, and wheelbarrow traffic, is one of the last of the outer villages that used to populate the North Coast. This chapter investigates how port noise came to index the place attachments and historical sensibilities of its settler residents, and how these attachments depended on dispensations the residents had long sought to disavow. In the early 2000s, the North Coast saw a number of noise abatement campaigns directed against the Rupert Port. What made the Dodge Cove case unique was not the force of its critique or the acumen of its criticisms. Rather, it was the position from which the critique was projected, one which foregrounded the complex symbolic politics of noise. Here, I will make two arguments. The first is for a particular conception of settler colonial listening. Settler colonialism is an affective and ideological system capable of eliciting powerfully felt relations to sound, I will suggest. What is notable about this system is how it can simultaneously focus listening and instruct subjects into recognizing their sensory ordering acts as insensible (e.g., as containing no trace of intentionality or focus). This distribution of sensing/nonsensing is neither smoothly operative nor guaranteed. It is selective, messy, and at times contradictory. This is indeed the case with settler colonialism generally, which is not a critique my interlocutors would have expected me to associate with their struggle. But settler colonialism is essential to consider if we are to grasp the limits and the challenges of anti-noise politics in Dodge Cove.

Specifically, I develop this line of argument by drawing on Mishuana Goeman's "settler grammar of place."[4] What Goeman's concept helps us understand is that settler colonialism need not appear as the possessive land-based project that it is. Rather, it can be delivered through a deeply aestheticized, feelingful, and ecologically attuned sense of place. In Dodge Cove, sense of place will link listening and noise to the faded symbolisms of postwar unionism, rich primary commodities economies, the back-to-the-land movement, and powerful ideas of nature, perhaps most of all. One community member told me that Dodge Cove is a place where people find "the natural rhythms of the coast." These rhythms are every bit as historical as they are natural, as I hope to show here.

The second argument pertains to logistical capitalism and the way its aural symbolisms are distinct from, but reinforcing of, settler colonial ones. In Dodge Cove, acoustically mediated presentiments of ecological loss will be energized not simply by the assumptions and anxieties constitutive of settler claims to place, but by an encroaching spatial envelope of neoliberal port development. "Port noise" was, for Dodge Cove, the leitmotif of the historical process they sought to arrest. I draw on Henri Lefebvre to show how port noise and the proliferation of logistical nonplaces that produces it are the direct consequence of the spatial strategies of the capitalist state.[5] Dodge Cove residents intuitively seemed to understand this, at least in part. For them, the transformations taking shape at the Fairview Terminal, along with the expanded cargo vessel activity, were audible reworkings of "state space"—that is, "homogeneous, logistical, optico-geometrical, quantitative space."[6] They symbolized "the port claiming more land as its own," to quote one resident. In pursuing noise abatement against the port, residents sought a "lived revolt" against the abstractions of economic globalization.[7] But in their opposition to the 24/7 barrage of beeps and hums they remained dependent on problematic grammars of place attachment. As I explore in this chapter, their resulting politics could not enact a bulwark against capitalist anesthetics. Instead, "port noise" would reinforce a "settler grammar of place" that was itself constitutive of capital's spatializing moves. Within a community avowedly critical of port development, carefully attuned to signs of environmental change, and still capable of recalling the intercultural fishing alliances that once supported communities up and down the coast, this is a troubling development indeed.

Grammars of Place, State Space, and Noise

As Jodi Byrd notes, settler nation-states like Canada have historically conceived of Indigenous peoples as populations that can be "made to move."[8] They can be made to move, and removed, because settlers will identify lands to which they have "come to stay." Beginning in the late 1960s, remote regions across North America saw new waves of settlement from white North American middle classes, disaffected with postwar life. "To a privileged generation exhausted by shouting NO to every aspect of the . . . society they were raised to inherit," writes Kate Daloz, in a history of this movement, "rural life represented a way to say yes."[9] The white North American residents who first settled Dodge Cove in the 1960s displaced the Norwegian boatbuilding families who had arrived in the 1910s and '20s. For them, the "yes" of "rural life" was not merely about new economic opportunity but the self-actualizing powers of nature. It had to do with modeled relations that were both revealing and obfuscating of the lands and waters where rural life was to take shape. These relations imply rules, indexes, and forms of imaginative spacing—in other words, a grammar. "Foundational to the normal modes of settler colonialism," Mishuana Goeman writes, "are repetitive practices of everyday life that give settler place meaning and structure."[10] Grammars point to highlights and points of emphasis. The circumstances I explored over two winters and three summers—a Dodge Cove of lush summer gardens, abundant berry-picking, and old wood boatyards—offered several highlights and points of emphasis, constitutive of what acoustical ecologists call *soundmarks*. But gaps and absences also matter to a grammar. Jodi Byrd develops the concept of "colonial agnosia" to explain how settler colonialism can be both a pervasive and "not comprehended" reality for those who benefit from it.[11] What became apparent to me over my time in Dodge Cove was that a pervasive sensitivity to industrial port sounds, to noise, also involved Byrd's "not comprehended." The comprehension of noise was also about, as Manu Vimalassery and colleagues summarize, "an investment in maintaining the failure to comprehend."[12] But let me say more about the other piece of this.

The story of Dodge Cove shows how preexisting patterns of "spatial violence" can be reanimated through the mediating force of noise and the interpretations it solicits.[13] The force of this association is made clearer by attending to one process in particular: state space. As Neil Brenner points out, Lefebvre used state space to consider strategies through which states seek to extend capitalist social relations onto previously marginalized zones within a national territory.[14] State space is also colonized space; it is space that is coproduced with settler colonial logics, if also a mutating, changing formation.[15] Beginning in the early 2000s, new processes connected to neoliberalism, port expansion, and global production networks began to redefine economic life on the North Coast. In anticipation of industrial development, new roads were built, formerly public spaces were fenced off, and a spate of new security measures began materializing. The audible dimensions of this onset phase of development would manifest to residents both qualitatively and quantitatively: more sounds, more times of the day, louder than had been typical, and with less association with "whatever's going on that we can see over there"—to quote resident Jeremiah Randall. For Jeremiah, port noise was "the first thing we felt in Dodge Cove" connected to development. Abstracted from local ways of life, port noise was a harbinger of a new economy. In Dodge Cove, the new economy posed a new relation to state space; one threatening the place they had held apart from it.

Port noise was thus interpreted as a sensible expression of something quite significant. It was a means through which residents could grasp the entailments of a port expansion: the novel forms of measuring, mapping, occupying, and controlling that come with expansions of state space.[16] As Lefebvre keenly recognized, a characteristic feature of state spatial transformations is the mobilization of new technocratic knowledges. Nation-states can utilize a flurry of technocratic procedures for addressing noise concerns: acoustical expertise, environmental impact assessments (EAS), liberal consensus discourse, open data platforms, and more. To understand how these procedures would achieve their effects, we need to link not only to state space but also to settler grammars of place. Analytic attention to specific places and specific historical understandings of sound and place is also needed. Ultimately, my goal in assembling these materials is not to criticize an antidevelopment community politics for its failings. Rather, it is to better understand how a community became mired inside a contradiction that delimited its appeals to both space and place. We need to recognize the impasse that communities like Dodge Cove would find themselves in, and then to keep going, in the hopes that we find novel narratives of struggle and perseverance, free from spurious senses of place, and the "sense of . . . complete U.S. or Canadian conquest."[17]

"Boat People"

"We have had, and continue to have, a number of concerns about development, but the noise was the first thing we felt. And the noise has been with us for years now." I met Lou Allison during one my first visits to Dodge Cove, and over the ensuing years we probably spoke more about the village than I would with anyone else. Lou came to the coast in the 1970s as a Gumboot Girl, a movement that drew hundreds of white middle-class women to the remote corners of North America in search of adventure and feminist self-reliance.[18] After living on Haida Gwaii for more than decade, in 1985 Lou and her family relocated to Dodge Cove. "We're boat people," she told me, as we sat on her back deck overlooking Hospital Island, the houses of Prince Rupert peering out from either side. "And I might be the only librarian in the North Coast who commutes to work aboard a skiff."

Lou and I talked a lot about books. I often saw her during her shifts at the Prince Rupert Library, where I liked to write. One of Lou's favorite books, she said one day, is *The Curve of Time* (1961), M. Wylie Blanchet's account of sailing the remote West Coast with her children. *The Curve of Time* is something of a regional classic. It is often read as a story of motherhood in postwar British Columbia, but for the purposes of my research it contained other insights as well. Among its notable details is the sensitivity of the protagonist, ever on alert for signs of danger in a beautiful but foreign land. "There is always the same kind of peculiar silence about all these old villages," the narrator explains at one point, in a chapter titled "Indian Villages." "It is hard to explain unless you have felt it."[19] This silence is a source of narrative anxiety, or so it would seem, and at one point the narrator seeks to resolve it by moving an Indian copper bracelet from its resting place. This provokes an audible remonstrance from an invisible source: "'Tch . . . tch . . . tch,' said a voice in unbelieving tones. . . . It was repeated in all the trees, on all the

FIGURE 3.1. Sarah Brown's house in Dodge Cove, Fairview Terminal in background, 2015. Photo by the author.

branches, from all the boxes."[20] I agreed with Lou that *The Curve of Time* is a portrait of the "rhythms of oceanside life." Its prose captures "what boat people do" as they confront the wonders and mysteries of the BC Coast. But another way to read it—the one to which I would turn—is as an allegory of settler disorientation. The book invites connections with the changing sense of place I would discern in Dodge Cove—the wide vistas and cool ocean winds, the unexpected energies humming within.

For Lou, as for many Dodge Cove residents, living in the North Coast is closely articulated with a sense of ecological attunement. Here, personal freedoms find form as palpably sensory freedoms: fresh air, quiet mornings, the slap of the boat on the water. Being "boat people" today means being able to explore and experience the North Coast, much like Blanchet had in her boat decades earlier. Birdsong, the soft rolling of the tides, and tapping corking mallets would be described by Lou's neighbor John Leakey as "part of the fabric." "The crows and loons!" another resident, Sarah Brown, exclaimed in response to my prompt. We had just finished a walk up CBC Trail. I had brought along my recording gear, and we had spent time stopping at select locations to listen to trees, creeks, the sky overhead. Sarah was one of the youngest residents in the cove. She had recently moved back from Prince Rupert to homeschool her five-year-old. Sarah was a talented artist and a consistent, enthusiastic interlocutor. "Those are definitely winter sounds," she noted of the birds. Ellen Marsh agreed. "But the noise in the area really affects how much they call," she added. Birdsong was a beloved local feature. It furnished the cycle of sounds that acted as a local orienting device—robins in the spring; sandpipers, jays, and herons in summer; loons, ducks, seagulls, and geese in fall. Ravens and eagles all the time.

As Andra McCartney suggests, soundwalks can be powerful ways of documenting attachments to seasonal rhythms.[21] We often can't recall the sounds that matter to us in absentia. What's needed is the correspondence of place, notes McCartney, with the processual acts of the soundwalk animating associations that are otherwise unavailable. From the half dozen soundwalks I led with residents in Dodge Cove, it was evident that a strong sense of place had been actively if not consciously nourished. It had been nourished by practiced local investments in walking, hiking, and boating through the coastal ecology. Birdsong elicited audible exclamation points that spoke to sense of place. Occasionally, birdsong also revealed its gaps—in that some calls were deemed not part of the fabric (the "invader species" Eurasian ring-necked doves, for instance). But overall, residents spoke of their ecological connections with great satisfaction. "I choose to live in Dodge Cove because it is rural living, close to nature, that has quiet, fresh air, clean water, food harvesting and hunting and fishing," Tommy Spiller told a local journalist in 2015.[22] Like Tommy, John Leakey's sense of place had been spatialized through years of local work. The audible relations he reenacted to me—whistling the songs he used when he hiked the bush ("so the bears know I'm here"); identifying returning residents by the sounds of their engines ("Vroom! Wub wub wub . . .")—suggested a range of place attunements. The sound of "the herring pumps" is the sound of "March, April, May," John explained, "because it follows herring spawning season." Corking activity (*pop pop pop*) happens in "the nine to five." We spent a long time talking about cycles: auditory, ecological, historical. John was one of the longest-running residents in the cove, having arrived in 1971. He hoped younger families—like Sarah's—would once again fill the schoolhouse that had been dormant since 2010. His sense of place sought its structure in this recurrence.

In their 1989 Official Community Plan (OCP), Dodge Cove's residents had endowed these coastal relationships with a name: "rural marine." The OCP, coauthored by many of the people I would later interview, discusses "quiet slow-paced living on large lots having easy access and a view to salt water."[23] I would also find this rural marine in a more recent document: the sound questionnaire I circulated across the village in fall 2013. It asked residents to describe the sounds of the village prior to the Fairview expansion.[24] The results offered an unexpected finding, not suggested in the community plan: in effect, rural marine can be very noisy. This came through in several reported details: the old Wahl family sawmill, humming away; the clanging interior of the working boat sheds; the *tap tap tap* of the corking mallets; the revving of power saws echoing across the bay. "When you hear outboards," one respondent wrote, "you grab your binoculars to see who's doing what." "The sound of the chainsaw," wrote another, "that's the sound of fall, ready with firewood, that is a *soothing sound*!"

Settler grammars of place are often marred by inconsistencies and incongruities, Goeman notes. Declarations of noise, meanwhile, are often anything but straightforward in implication—a point sound studies scholars have been at pains to observe.[25] The combination of the two, I would come to learn, supported a range of interpretative possibilities: certain sounds, at certain volumes: OK. During certain times of the day: tolerable or even desirable. In other contexts, unequivocally not. Given all of this, what was striking about the responses my sound questionnaire gathered was the consistency. It featured two claims: first, that the noise coming from Fairview Terminal was more than unwanted sound.[26] Rather, it symbolized the arrival of a historical process that would negate the spatial freedoms people in Dodge Cove had come to expect. Noise was a symbolic force carrying material changes. Otherwise known to some as Fairview's "24/7 anonymous sounds," noise symbolized the abstractions of a changing economy: a port terminal managed by a quasi-public port authority, later linked to Asia-Pacific trade but administered by a New Jersey company (Maher Terminal) and, later, a Dubai-based one (DP World). People wrote about this idea, and they talked about it. They formed connections. Given that the harbor on the Rupert side of Chatham Sound was being slated as another expansion site, noise "one day means that our boats will have nowhere to dock," Doug Bodnar explained. "Port noise" was set against the noise of the rural marine that had existed here before. The "local noise" was "part of the fabric," John Leakey told me. "There is a big difference between the noise we heard before and what we hear now," Carol Brown, a retired schoolteacher, explained. We were sitting around her kitchen table, drinking coffee. "Noise was seasonal then. You could set your clock to it if you needed to." Lou concurred. "There was a lot of boat noise, back in the day," she admitted, "but it was a different kind of noise than this. We knew who was making the noise." Added Carol, "But no one complained about the noise back then. It stopped when it needed to."

The second claim was that industrial noise has been a regular feature of life in the Prince Rupert harbor area for much of the twentieth century. The tail end of BC's "long boom" (i.e., the 1960s to early 1970s) resounded across Chatham Sound in the sounds of fishing, logging, and all the machinery needed to process them. Throughout those years, the canneries—North Pacific Cannery, Babcock Fish Plant, Atlin Fish Plant—were still teeming with people. "Oh definitely, the waterfront was a going place," one Prince Rupert resident told me. "There was lots of shunting. Lots of fish brought in and exported by rail car. Northland Steamship docks, long-shore noises . . ." "There were train horns, ferry horn blasts, and even tugboats [to] liven things up!" recalled another. Residents working the dock on a busy Monday would have heard the commotion of boxes being moved, the cacophony of different voices, maybe even the rippling clatter of ice being thrown into metal train cars to keep fish cold for continental travel.[27]

When this soundscape changed is difficult to pinpoint. It was certainly different by the mid-1980s, when the rationalization and decentralization of a once-mighty halibut industry occurred.[28] The volume of trains coming to the waterfront by this time would have lessened as increased competition from low-cost production regions led to cannery closures and firm amalgamations. Regional outmigration, something that remains a problem for the North Coast industrial economy today, began to take effect around this time. By the mid-1990s, a new regional sensibility about the waterfront began to find expression in local writing. "That whole section of waterfront overgrown with wildflowers and weeds is like a huge graveyard of what used to be a busy spot," Phyllis Bowman observed in 1997, "—with dozens of boxcars being shunted about by puffing engines and hundreds of workers busy at their chores in the ships and roundhouse and stations, with boats of all sizes and descriptions coming and going at the long docks 24 hours a day, every day."[29] A 1996 epigraph by Ray Gardiner in the *Prince Rupert Daily News* captured a new nostalgia: "I fondly remember as a child on a cold winter's night listening to a train recede into the distance. I tried many times to stay awake until the sound completely faded. I don't think I ever succeeded."[30]

Owing to its pattern of long-standing New Democratic Party (NDP) leadership, neoliberalism's entrée into the North Coast began later than it would elsewhere in Canada.[31] In was only by the mid-1990s, when the political fortunes in Prince Rupert and the surrounding regions had decidedly turned for the worse, that neoliberalism's market rationalities became dominant. With the 1970s population halved and the slate of primary commodities economies struggling, the provincial government saw fit to unleash a wave of cutbacks. Social welfare programs and labor protections were slashed. Businesses relocated. Fishing used to be "fast and furious in Dodge Cove," Bill Smith told me as we sat on his porch eating peanuts. By this point (it was spring 2015), the village had only four commercial fishermen left—"just enough to ensure the dock was maintained by the federal government," he added. Bill was a gill net fisherman. He had spent much of the previous decade working, grudgingly, for Jimmy Pattison, a Vancouver property magnate who has long been held as a symbol of fishing's demise in the North Coast (his Canadian Fisheries Company was one of the last operations in the North Coast in the early 2000s and has since moved). Originally from West Texas, Bill said he missed little about the lands of his birth except the country music (the George Jones album playing in the background during two of our interviews lent this statement some credibility). Whatever his Canadian experience, Bill sounded, to me at least, like a Texan. "Fishing has ruined my ears—why in the heck you want to talk to me!" he roared at me in his drawl one day. To help him deal with the tinnitus, Bill had radios constantly playing at nearly all hours. The very notion, he reflected to me at one point, still made him laugh out loud.

It was Bill who first appraised me of the jurisdictional challenges port noise had raised among the community. Already suggestive of the region's changed industrial footprint, port noise additionally revealed the flaws in two recent municipal planning decisions.[32] The first decision pertained to delays around proposed changes to the Dodge Cove OCP. With community support, the Skeena Regional District had hired a planning firm, Urban Systems, to develop an updated document that would better recognize the community's novel development concerns. Among other things, the plan would have established "community buffers" (in the words of Doug Bodnar) around acceptable industrial noise levels and other environmental contaminants (e.g., light, smoke). But in 2016, BC Minister for Community Affairs Peter Fassbender issued a ministerial order that the Dodge Cove OCP would not be adopted by the provincial government, as it was not in line with the BC Oil and Gas Commission.[33] The OCP decision came on the heels of another decision, from a year earlier. In 2015, the BC Province had quietly rezoned Digby Island (on which Dodge Cove sits) as industrial, informing neither the village nor the regional district of the change. The waterfront lots in the unincorporated village of Dodge Cove are under federal jurisdiction as Crown Land, but the region's intertidal areas are administered by the port, pursuant to the authorities granted by the Canada Marine Act. Bill and I spoke just as word of the rezoning leaked but before the revised OCP was blocked. Still, villagers were already seeing the two as effectively leaving them "bufferless" (in Doug's words) from industry. They were concerned that the port, acting on behalf of the development it championed, might demand further accommodations from the community across the harbor.

In Lefebvrian terms, the changes Bill described were modulations and reconfigurations of state space. And like others in Dodge Cove, Bill consistently raised his industrial development concerns to state space authorities, in this case City Hall. At the same time, Bill fully recognized that the city's jurisdictional limits precluded it from acting on most development-related matters. He also knew that the port was not the only actor supporting the changes. The rezoning and blocking of the revised OCP was also supported by interests in the Aurora LNG project, which was to be located on the west side of Digby Island. "The seasons keep getting shorter for commercial fishermen in the region," Bill noted. "The noise problem is especially frustrating for us because industry is being given a free pass, on top of its threats to our fishing."[34] Across the harbor, previously unused sections of the Rupert waterfront were by this point being reclaimed by Canadian National (CN) Rail in anticipation of expanded trade volumes. For Bill, the preferential treatment was there for all to see and hear.

Bill's neighbor, Des Nobels, would expand on these relationships. "Rupert is encircled by industry now," he told me. "You would not believe the degree of intimidation that comes onto people who speak out against the port." If Dodge Cove had a political leader, it was Des: a wiry man with a graying ponytail and a sharp demeanor. Des was an environmentalist, but not of the sort I had encountered at Cetacea Lab. His animal of choice was salmon, not whales, and his approach was less mollifying and more antagonistic. In the late 1990s, he became known on the Coast as Black Des for his leading role in the blockading of Alaska Ferries. Since 2007, when he retired from commercial fishing, Des has played a steering role in T. Buck Suzuki's habitat preservation efforts and worked to empower local fishing unions. When I met him several years later, he was also serving as an elected representative for the North Coast Regional District, making appearances at industry open houses, and cultivating his considerable talents for silversmithing. He was, by all accounts, a committed local.

Speaking to Des about my research felt vindicating. He was visibly annoyed by environmentalism's new obsessions with social media. He expressed disdain for the "foreign-owned megavessels" issuing ocean noise and was quick to celebrate the defeat of the Enbridge Northern Gateway in 2016 for its "coast-wide" benefits. Most of all, Des loathed the idea of a toothless regulatory government, a situation for which he blamed a singular local adversary. The port was cynically using Prince Rupert's unemployment problem to grow rich, he exclaimed, exerting its influence on city council while denying any responsibility for addressing the city's crumbling infrastructure.[35] "This community bent over backwards to help the port get off the ground," he continued. "And now that they are well-established, it's 'We are the kings!,' 'How *dare* you question us?'" Des wanted me to examine the "broken system" through which noise was being addressed by all levels of government. Environmental assessment (EA) has acquired an "overbearing prominence" in the daily lives of Dodge Cove residents, he added. If people got a first sense of the government problem through listening to noise, they received a second sense by learning how the state factualized, codified, and misrepresented this noise through EA. Though he never used terms like *neoliberalism* or *state space*, it was clear that Des saw a unifying logic: the policies guiding port restructuring were part of coordinated state-capitalist strategy to alter the logics of governance in accordance with the needs of the market. Through noise and other things, development was actively being pushed into his ocean-view backyard.

In the spring of 2014, I met with a small group of residents from Dodge Cove and Prince Rupert to take on the Fairview Terminal's environmental screening document (ESD).[36] By EA standards, the three-hundred-page ESD was light reading. In struggling through it together, I hoped we might challenge its compartmentalizing logics—including those which divided general concerns over developments into a patchwork of projects. Fairview's ESD had noted that "noise related to train traffic will increase by 50% compared to previous levels," I read aloud at one point.[37] But as Prince Rupert resident Brian Denton fumed in response, train whistles already made his life "absolutely unbearable, far more than the '50%' number suggests!" Brian had elaborated on this point in a letter he penned to Health Canada a few years earlier: "My wife and I have lived in our same location in Prince Rupert for close to 40 years and we can state categorically that the noise from rail traffic for the past 5 years has been 1,000 times worse than ever before."[38] I took note of the fact that the Dodge Cove attendees had little to say about the trains that upset Brian and his neighbors. The "crane-humming and beeping sounds" that upset Dodge Cove residents Doug Bodnar and Ellen Marsh appeared in the Fairview EA as "noise"; the train whistling that upset Brian Denton also appeared as "noise." But as I came to further understand, neither the Dodge Cove nor the Prince Rupert residents were particularly interested with the losses that noise posed to those on the other side of the harbor.

Another common theme was apparent. On both sides of the harbor, state space was becoming more alienated from its users, not only by virtue of its appearance in reifying technical discourses, but likewise through its capacity to partition understandings of where community's interests began and ended. This was apparent in the discussions we had that day about the ESD. It was also apparent in follow-up discussions with individuals. Confused appeals to state regulators seemed only to make the divisions worse. By the middle of my fieldwork, I had amassed enough comments and newspaper clippings to construct a script of the frustrations being routinized in responses to noise:

Resident (email to me, August 10, 2013): When do I get to make a "comment"? Why can't I make a comment now, the noise is already a huge fucking problem?!

PRPA representative (email to Resident, August 12, 2013): Once the assessment begins, public consultation plays an important role. You will be invited to learn more about the proposal and ask questions of the proponent. . . . You will be provided with the opportunity to provide feedback.

Resident (email to me, February 13, 2013): Hey Max, I wrote to CN about the trains and then they wrote back saying I should write the proponent. Who is the proponent? Is that Pinnacle or the Port?

Industrial proponent (then Pinnacle Renewable Energy CEO Leroy Reitsma, quoted in *Northern View*, May 22, 2012): We're now getting into public comment period. Or getting close anyway, and I would say that through the next month we will be looking to try and engage the public around the design of the terminal and share [a] lot of information about the design and what we've done to reduce dust emissions, reduce noise and reduce the visual impact.

Resident (field notes, December 9, 2014): I'm not going through another one of those damn things again. It's just another rubber stamp. Forget it!

Dodge Cove residents were particularly aggrieved by the "clumsy" regulatory process. "If you must allow gigantic and incredibly noisy generators on Digby," one resident posted in response to the Aurora LNG proposal, "at least have the decency to build them well into the old artillery hill . . . to minimize the destructive influence the constant noise will have on our sanity, sense of community, and private property investment."[39] In his opinion piece in the *Northern View* (February 18, 2015), Tommy Spiller echoed this sense of disappointment: "I'm now 63 years old and think that I should be able to retire in the fashion I've envisioned for many years. When the terminal was in the planning stage, we were assured the noise levels in operation would be kept to an acceptable minimum . . . lies." A cycle had been established whereby Dodge Cove residents sent impassioned and informed letters to the online portals on the EA project websites. Receiving anonymous replies, they dialed phone numbers that government representatives eventually answered, only to be directed back to the portals. One day, Lou gave me the minutes from the November 7, 2014, community meeting in connection with the EA for Aurora LNG. In it, I found a lucid account of the impasse that the community was confronting, as offered by one exasperated resident:

> I have the project description, there is a lot of potential waste and contamination, and there is a whole page of species that could be affected, seven that are endangered or species at risk. There are hydrocarbons and greenhouse gases that will be emitted, but you don't discuss how much. In regard, to human health, your contaminants will be SO_2 and will mean a degradation of air quality, sensory disturbance, social effects, sensory effects, cultural effects, and mean changes in traditional use areas. My one question is, what is it that you are asking us? Are you asking us to say yes to all of that? It is coming in one quantity or another, we could argue about any greenhouse gases or how much noise, we heard the noise this summer. Are you asking how much disturbance we will put up with in our community?

Community Fissures

In a developing port city, responses to port noise would be mediated by, and work to reinforce, a fractious urban politics. Noise also revealed some of the underlying commitments groups in seeming opposition held. In *Pipeline Populism*, Kai Bosworth considers the possessive frameworks that can inform settler oppositions to development. The "territorialized resentment" Bosworth identifies helps to clarify how investments in private property might have been working to mitigate against community solidarities on the coast.[40] I found evidence for such resentment in a now-defunct chatroom called Hacking the Mainframe—a community forum that existed in Prince Rupert in 2007–10 before being superseded by Facebook. On it, anonymous Prince Rupert residents castigated Dodge Cove residents for an antidevelopment politics they read as sanctimonious and self-interested.[41]

Independent (February 14, 2008): Funny how people think, no whistles for three years, that's what happens when there is no work, now there's work, trains are back, think back people Prince Rupert is here because of the railway, progress sometimes has drawbacks but that's life.

bubbasteve735 (July 22, 2010): First, they complain that there's nothing going on in this town, next, when there is something going on, and bringing money into the town, they complain about the noise. . . . Suck it up cupcake, if you don't like it . . . MOVE. I don't hear those noises on my end of town.[42]

Dodge Cove residents struggled to understand the vitriol. "We're not against growth, but we don't want our nature ruined," exclaimed Lou Allison. "I don't see what's so controversial about that." For Lou, the chatroom dialogues spoke to a Prince Rupert that had become increasingly "divisive." On Hacking the Mainframe, and later Facebook, some people celebrated port noise as a "price of progress." How, I wondered, were they being informed about the other experiences such progress entailed? If the community center or the local pub had once provided space for dialogue, by the time of my fieldwork these spaces had become significantly reduced. Since 2008, when wireless internet services became available, social engagement in the North Coast has increasingly moved to online social networks. The ubiquitous computer glow in the windows near my Prince Rupert apartment seemed to symbolize the resulting diminishment of public space. But it was the loss of fishing that mattered most. For residents like Des, the city's once-vibrant union culture had been the source of a truly collective resistance to government and corporate whims. John Leaky was disconsolate: "We're way past the time of alliances, Max. We've got new generations that have to make those connections, now." For Leaky, the flipside to Bosworth's "territorialized resentment" was territorialized grief. The "current of libertarianism" political scientists Gary Wilson and Tracy Summerville observe across northern British Columbia had come to settle here as an elegiac isolationism: bounded and wary.[43] To invoke the language of Wendy Brown, place attachments had fused with a community's "wounded attachments"—if less focused on the state, in this case, a sensibility nevertheless deeply preoccupied with what the state was now denying.[44]

After one of the last soundwalks I did (in December 2016), a Dodge Cove resident named Wendy Brooks asked us all to stop at a grassy patch beside the schoolhouse. "There's a great blue heron rookery in those trees over there," she noted. "Or was. Because I am not sure they are still there. I haven't heard them calling for months." The others nodded, and an uneasy silence passed. I later found out why. Wendy's theory—which attributed the disappearance to rising levels of noise—had become a source of some local controversy. She had published an op-ed about the herons in the *Northern View* and had written extensive details in the Aurora LNG comments section. She would speak about it to anyone and everyone in town. But as one resident explained, neither Aurora LNG nor state biologists who had investigated the herons had come to her conclusions. People were worried "that [they would] be perceived as cooking the books" in trying "to make a connection that wasn't there," one resident, preferring to be anonymous, explained.

The power of birdsong to convey nature's general communicability, to tell us important things about the state of world, can define an array of sense makings.[45] In Hartley Bay, elders trace the ripeness of the salmonberry in the melody of the Swainson's thrush (*Simiik'isk*)—picking the fruit only when the sound is right. Widespread avian declines have given birdsong a central place in many "anxious ecologies" of the Anthropocene, with sensings of extinctions spanning new urban spaces and communities of listenership.[46] Wendy's comments and the concerns about them made me wonder about birdsong's symbolic capacities. On the one hand, the purported link seemed an over-easy attribution, as in the way several residents had discussed Eurasian ring-necked doves as "coming from China." And yet, Wendy may well have been right. As I write these words, Dodge Cove remains without a heron rookery. Industrial development might well have been the cause of their relocation. What the episode revealed were small but significant cracks in a community's self-conception, in its sensorium, and in the nature it revered and the grammar mediating its expression. The silencing of the herons, and the silencing of a discourse connected to them, spoke to something meaningful: a community less able to cultivate a shared sense of place. But what was, in fact, the logic through which its grammar was seeking to operate?

Unsettled Natures

Buried in the transcripts of one of the many community hearings about the Aurora LNG project, I found a document discussing the history of Dodge Cove. According to the notes it contains, the following passage was read aloud by Des Nobels to an energy proponent representative, in order to communicate the connections that would be lost with the project:

> It was founded by non–First Nation and First Nation people who had married. . . . The First Nation connection is significant. The First Nations woman [*sic*] who married the non–First Nations men passed on that knowledge and access to the land and the community has availed itself of that over the years. When we say our Island, we think of it as our community, there is no separation. . . . We have carried on with that First Nation connection, even though we are not First Nations, as we think it's a good way to live.[47]

It is only in the telling and retelling of stories, in the life lived and recalled, that space becomes place, that sense of place acquires its grammar and meaning. What this passage acutely reveals is how settler grammars of place are historical constructs. They are predicated on the ability to hear a past discontinuous from a present. In Des's account, Dodge Cove's entitlements to the land—its demarcated spaces and natural contents—flow from a now-vanquished Indigeneity. The land had passed to them in the ur-progression that began with contact. Even as the story hesitates at the analogy ("we are not First Nations"), it doubles down at the next moment ("we think it's a good way to live"). Herein lies the tension that structures the community's grammar of place; its enduring sense of a proximal Indigenous past that nourishes present-day community life but cannot be fully heard in the present.

This grammar is by no means unique to Dodge Cove. It is the hidden side of the entire back-to-the-land movement that commenced rural resettlements across North America in the early 1970s. As a formation, this grammar has evolved and taken on different features in different places. In Dodge Cove, it fuses with the creeping isolationism I found in the OCP, which celebrates the rural marine while worrying that improved boat access would result in an "unwelcome increase in non-resident visitors" and loss of "community atmosphere."[48] It is above all "possessiveness," as Dylan Robinson incisively notes, that allows settler colonialism to construct its line through history; to miss the gaps, reversals, and alternatives that other listenings can reveal.[49] In the North Coast, this listening can be linked to a range of formations, including the acoustic ecology movement that was forming in the Vancouver Lower Mainland during the same period of the village's resettling. Des wasn't familiar with acoustic ecologists Bruce Davis and Peter Huse, who had worked up and down the North Coast through the early 1970s, documenting the fading elements of an interpreted Canadian experience. Nor had he heard of influential BC folklorist Imbert Orchard.

In his popular 1970s CBC radio show, *Between Ourselves*, Orchard recounts a visit to Kitselas Canyon on the Skeena River and its old village sites. Much as in Des's story, Orchard fondly places Indigeneity, but even as he hears it, Indigenous people are nowhere to be found ("The voices of Kitselas can still be heard, and there are many of them," Orchard says, before enacting his telling temporal cut: "It is still and always will be a place of legend and story"). It was no accident, the historian Robert Budd notes, that Orchard had seized on Boasian notions of salvage to inform his work. His goal, like the many other settlers who came north with their recorders and assumptions of history, was to possess sounds before progress washed them away.[50] In 2015, Des could also hear progress. He wanted to protect his community from its effects. But in support of this goal, he imagined a community that came into existence through available grammars of place.

Periodically throughout the year, Indigenous cultural activity asserts itself along Prince Rupert's waterfront. Young people and elders pour down from the outer villages. To celebrate the springtime salmon runs, Mariner's Park becomes a chorus of drumming, singing, and feasts. Dodge Cove residents partake in these events. They tout the intercultural learning opportunities and festivities. These facts do not disprove my argument. My concern is not with whether or not community members care about their Indigenous neighbors—something I do not pretend to have authority to address. My concern is with a settler grammar of place as a historical fact; as a formation with durable traces and sustaining investments "in maintaining the failure to comprehend."[51] This maintenance concerns possession and what communities are and are not willing to possess in the form of land. This erects challenges in what a grammar is willing to disclose. Even Blanchet's narrator in *The Curve of Time*, for all her perceptual acumen, cannot transcend its impasse: "Impossible to explain to *them* that I was trying to save their Past for them," she exclaims to the voice in "Indian Village."[52] And so she does what settlers have historically done, which is cultivate her sensibilities through a collective grammar that allows her to exist as such.

There is an interesting irony to the story of Dodge Cove's resettlement in the early 1970s. Just a few years before this, in the late 1960s, the North Coast Prehistory Project had excavated eleven major Indigenous archaeological sites around the present day village.[53] It remains one of the most significant discoveries of Indigenous occupation on the entire West Coast. Excavation work less than 500 meters from the Dodge Cove marina suggests Indigenous inhabitation stretching back over five thousand years. It suggests the existence of flat terraces large enough for rows of multiple plank houses; large wooden plank structures erected in rows; and copper artifacts, impressions of which would have remained visible as rectangular surface depressions for decades after.[54]

Bryn Letham is an archaeologist who has worked throughout the North Coast and extensively with the Gitga'at Nation. I first met him aboard the *Tsimshian Storm*, during one of the four-hour boat rides visitors and villagers take between Hartley Bay and Prince Rupert. At one point I asked Bryn what life in the precontact Dodge Cove area might have sounded like. I expected no response. But Bryn considered the spring migrations and the historic rainfall patterns, the recorded battles, and the rich archives of community songs. He began discussing the seashell middens archaeologists continue to uncover in various parts of the North Coast today: "We'd hear crunching as people walked over the shell terraces that people built their houses on to be drained from the rains," he wrote in a long email to me, "and we'd hear the clattering of batches of shells being thrown off the front of the sites to build up those terraces." I conveyed Bryn's story to a Gitk̲'a'ata friend named Clyde Ridley one evening, over dinner. Clyde replied that the clattering batches of shells had likely served as an alarm system for predators, such as wolves.[55] Prince Rupert's harbor once encompassed an entire network of Indigenous villages, full of audible relations like this—"tens of thousands of cubic meters of shell and other cultural debris, accumulated."[56] Neither Bryn nor I can say how these grammars were lived—are formed—by Indigenous peoples. There are other authorities for this. But as settlers, we are pressed to recognize how transiting waveforms can also create a false sense of a region's settledness; how they can sustain a grammar as "an extensive and constitutive living formation," even as places speak otherwise, perhaps in as mundane a sound as the crunching of seashells.[57]

Conclusion: A Doubled Sonic Allegory

In this chapter, I have presented the story of noise in Dodge Cove as an objective allegory of a doubled forgetting: of state space and the coloniality upon which it is erected. Hemmed in by neoliberal expansion and an elegiac place politics, port noise would not galvanize new political alliances but the protective tightening of a grammar of place.[58] To a certain extent, this story is also one of generational change—the diminished freedoms that come with aging; the disintegration and loss that marks all community life. But it should not be reduced to this, and nor should it be reduced to a moralism of settler colonial failure. One reason why, I hope to have shown, is because such a judgment would deny a more supple understanding of how settler grammars of place actually function: how their ecological attunements and ability to place rich memories give them considerable explanatory powers. These grammars are what allow new generations of settlers—like myself—to seek the continued promise of freedom on the North Coast. They are also what allowed me, in my Blundstone boots and wool sweater, to feel welcomed in Dodge Cove. Settler grammars are important to understand because history suggests they can be hard to undo. On the North Coast, they may well continue long after communities like Dodge Cove are gone.

It is also clear, however, that the challenges Dodge Cove confronted were not limited to colonial dynamics. They involved the vicissitudes and repressions of an ongoing period of spatial reorganization. In port noise, residents encountered many of the things they feared about development: a remote technocracy given to new forms of calculation, a government's continual prioritization of the market, an increasingly aggrieved local populace, a visceral encroachment of state space. This noise offered a reminder of what was being lost, including the sounds and the relations of a better community. Writing in the 1960s, Lefebvre hoped that a political awakening of the senses would give rise to new kinds of community. He called this process "autogestion" and proposed it might contribute to the rejection of state space. While not discounting his vision, the present story suggests that settler grammars of place can limit this project in significant ways. This was an idea Des imparted to me during one of our last meetings. "Your generation is in trouble," he said, as we stood waiting for the ferry on a wet spring morning. "There's no labor movement anymore. It's over." The words felt like a verdict; an opinion from an authority I respected.[59] There was little to do but wait. But even as we stood there, other grammars were at work around us, and in ways that challenged this verdict. I turn to one in chapter 4. It involves the emergence of other sounds on the North Coast, other histories, and new connections: a "cacophony" of Indigenous music.[60]

4

ANCESTRAL WAR HYMNS OPACITY AND INDIGENEITY IN GYIBAAW

"Whenever we play the music of Gyibaaw," Spencer Greening explains, "we truly believe we are bringing something in. An entity. Whenever we call on the Naxnox, we get shivers everywhere. Because it's a real spiritual thing: those spirits come and take a form and possess certain things." I am sitting in a sports bar in Prince George, BC, on a winter night in late January. Spencer's face—wide, beatific, grinning—is backlit by the televised hockey game. The two figures on either side—singer-songwriter Jeremy Pahl and guitarist Brandon Dyck—are bathed in ornate shadow. "The Gyibaaw was a Naxnox," Jeremy continues, referring to the Ts'msyen primordial power. He pauses, adding emphasis: "And still *is* a Naxnox." Gyibaaw the band formed when Jeremy and Spencer were seventeen. It has been broken up for more than a year at this point. But it seems to me that both the music and the spiritual being continue to manifest for Jeremy and Spencer as a presence, in the sense that Gyibaaw might be here, with us, in the room.

This chapter is about the power of musical sound and the music of two cousins from Hartley Bay. Echoing out from a remote corner of the North Coast, Gyibaaw spanned a historical period of immense change for Indigenous peoples: between a prime minister's 2009 declaration that "Canada has no history of colonialism" and the dramatic emergence of the Idle No More resistance movements three years later.[1] To consider the band's legacy in this span is to consider music's prefigurative possibilities, including its capacity to prefigure developments that redefined settler-Indigenous politics across North America. "We would also uncover the same transformations, the same progress, and the same eagerness if we enquired into the fields of dance, song, rituals, and traditional ceremonies," Fanon famously wrote in *The Wretched of the Earth*. "Well before the political or armed struggle, a careful observer could sense and feel in these arts the pulse of a fresh stimulus and the coming combat."[2]

But what is the nature of the relationship between the arts and colonial struggle? And what does it mean to further suggest, as Fanon does, that decolonization "infuses a new rhythm," along with a "new language and a new humanity"?[3] In a perceptive reading of Fanon, Glen Coulthard worries about "overly instrumental" interpretations of culture's role in decolonization.[4] His concern is instructive for the present case: While musical practices of self-affirmation may indeed be crucial to transits from colonial to decolonized subjectivities, is the path as straightforward as Fanon seems to imply? In Gyibaaw's case, it would seem that something else was going on—more akin, perhaps, to Gerald Vizenor's "active sense of presence," to capacities to establish connections with histories, places, and agencies in ways that eschew standard political logics.[5] This will be a challenging argument for me to make, and for reasons I go on to explore. (But at the outset, it is helpful to establish myself as an outsider to much that follows.) Since first hearing about them through a friend, I had grown immensely curious about Gyibaaw's music.[6] Long before meeting them, I had watched them play in the dozen or so live videos fans had posted on YouTube. I had read their liner notes on *Encyclopedium Metallum* and all the local music reviews. This led to conversations with friends, fans, and old bandmates; more for personal interest, at first, than anything to do with research. Eventually, I found a tape cassette of Gyibaaw's debut album. I blasted it on the shitty F150 speakers of the '86 Dodge Ram I had for fieldwork. It was only after *Ancestral War Hymns* became my de facto soundtrack to the highways of the North Coast that I began to ponder some of the questions it raised. The underlying momentum behind the sound and its "sense of presence" was something I needed to understand, even if its sense of inaccessibility never went away.

As we talked in Boston Pizza, Jeremy and Spencer seemed reluctant to engage in my associations with the coast with much detail. They were more interested in discussing some of the practical elements that had marked their band's career. It was apparent that the cousins were still struggling to come to terms with the cause of Gyibaaw's breakup. It had not been for a waning sense of Gyibaaw's power. If anything, it was the unexpected surplus of meanings their music had produced. From these came success and great personal discovery, but also a tragic and seemingly career-ending development: the embrace of Gyibaaw by neo-Nazis, who have long populated the fringes of extreme music in Canada's resource north.

The story I tell here is marked with ethnographic limits. It acknowledges my partial perspective on the Indigenous musical vision that was Gyibaaw. At the same time, it seeks to grasp the significance of a non-Indigenous position to the politics of Gyibaaw's music, and the fixities of race and "hungry listening" to Indigenous politics more generally. I start by outlining a theoretical framework that will assist me in this dual objective. Next, I consider the unlikely genre that served as the medium for the band's musical expression: black metal, a sound almost completely unfamiliar to the northern BC context where the band emerged. Gyibaaw was a red critique of the whiteness of black metal. It was also a celebration of a sound that has become vital to expressions of Indigeneity worldwide: a "noise uprising" that has echoed out during a period of heightened industrial despoliation of Indigenous lands, waters, and bodies.[7] Returning to Fanon, I will stress that there are multiple musical connections that are important to explore if we are to understand music's role in decolonial struggle and the politics of Indigeneity. Here, I will also seek the assistance of some key contemporary thinkers on Indigenous music, including Jarrett Martineau and Dylan Robinson. In the final section, I return to the North Coast, where I recount some stories of the band's eventful touring career. Throughout, I draw on conversations with the band's friends and relations, including Spencer and Jeremy's grandfather and musical inspiration: the country singer Johnny Pahl. Years later, Gyibaaw rings out as a visionary example of survivance, and Indigenous place attachment.[8] Or, to quote Jeremy, "Gyibaaw was about a homecoming."

Opacity, Fugitivity, and Indigenous Music

In their 2014 essay on Indigenous aesthetics, Jarrett Martineau and Eric Ritskes capture something of the impulse I want to explore in these pages. They argue that Indigenous art models what they call an "aesthetics of fugitivity": "not an abdication of contention and struggle," they write, "but a reorientation toward freedom in movement, against the limits of colonial knowing and sensing."[9] This idea chimes with much recent work in critical Indigenous theory, which stresses the ways in which Indigenous art expression can be "unintelligible to the western and/or imperial ear."[10] For my purposes, this interest in opacity establishes a tension with the prefigurative political moment Fanon points to and Coulthard nuances. From Fanon, we can listen to Gyibaaw for presentiments of the broader historical processes to which colonized peoples are subject; from Martineau and Ritskes, we can listen to observe a resistive dimension, one that is also important to their politics. There are two theories of musical mediation at work here, and two social realities. I want to propose a listening that does not dispute either, but works to press them into a dialogue—Indigenous black metal as a site of departure or flight, while simultaneously operating as a grounded cultural response to material terrains of politics. I am particularly interested, then, in the articulatory power of Gyibaaw's musical sound, one whose "coded articulations" supported a surplus of interpretative possibilities.[11] Although ultimately secondary to the personal and territorial connections the music established for Jeremy and Spencer, none of the interpretive possibilities Gyibaaw provided would be more significant to the band's career than would involve musical white supremacy, the neo-Nazi venerations that forced Gyibaaw's exit from the black metal scene.

As Spencer liked to remind me, the word *Gyibaaw* has no English correlate. The translated Sm'algyax word *Laxgibuu* describes one of the four constituent clans in a culture who once occupied territory in the Lowe Inlet area of Gitk̲'a'ata Territory. This comes to close to *Gyibaaw*'s implied sense, but the reference to clan structure does not capture another connotative meaning: Naxnox, roughly equating to a powerful supernatural spirit, or entity, integral to Ts'msyen culture. As a Naxnox, *Gyibaaw* is of a piece with powerful local cultures and figures, captured in myriad legends of the North Coast.[12] But *Gyibaaw* also extends from a third tributary: it is a ghost story, crafted and elaborated to Spencer and Jeremy by their grandfather, Johnny Pahl, over the cousins' extended childhood stays at his home in Prince Rupert. "He told us about the Gyibaaw at night," Jeremy explained to me, "to make us do good."

FIGURE 4.1. Gyibaaw, live at Little Big House, 2010. Photographer unknown.

What this combination of elements points to, when read alongside the "fugitive aesthetics" of Martineau and Ritskes, is the analytico-linguistic problem of opacity. As numerous musicologists have suggested, opacity is integral to the aesthetics of black metal.[13] It captures its mystifying Romanticist impulse: "absolute inwardness" turned into "dreams and visions."[14] But as Martineau and Ritskes recognize, opacity is also integral to the politics of Indigeneity, and in ways that help to reveal Gyibaaw's other cultural touchstones. Helpful in bridging these is a consideration of the setting within which Gyibaaw's music took shape. From one vantage, at least, opacity seems entirely apposite to the North Coast: an Indigenous landscape subjected to perhaps unparalleled colonial efforts to dissever music and language from place and peoples; to both know and to possess sound.[15] Sections of this ongoing history are well established: the salvage anthropology extending from such anthropological luminaries as Boas, Lévi-Strauss, and Barbeau; their various theoretical models for arresting Indigenous sounds and aligning them with formal linguistic indexes and classifications; the support always offered by the Canadian state. Across this span, salvage anthropology betrays an excessive need to make sound legible, as when Barbeau proposed the musical staff as a rack "upon which to pin down [Indigenous] sounds and rhythms."[16]

Against this legacy, opacity would seem to present an apt aesthetic response. Performatively, if not otherwise, it would allow Indigenous musicians to elide the logics of capture that have marked Indigenous experiences of modernity for hundreds of years. But in order to recognize the value of Martineau and Ritskes' critique, more needs to be said about the ways opacity might interact with musical conceptions of place, and the sounds Gyibaaw's Jeremy and Spencer would solicit in and through Gitk̲'a'ata. Spencer told me that the exact moment he played Gyibaaw, a multiplicity of objects, beings, and places fused in expressive relationship. "You open up your spiritual body with the land that you reside in," he told me; "the spiritual body of that land can then access you." To the outsider, it surely seems remarkable that such an exchange is not only possible, but possible through the music of black metal. How might we begin to explain it?

One the one hand, Gyibaaw's opacity will suggest the temporalities of an Indigenous North Coast, with connections that move "beyond settler time."[17] On the other, it will present a set of cultural materials that comprise a powerfully contemporary musical project. Here, opacity helps to disclose a negotiated participation in globalized circuits of culture and resurgent Indigenous movements that both desire recognition and refuse its mainstream presentations. In his conception of culturally expressive "moments," Stuart Hall describes "a fusion of contradictory forces . . . a 'condensation of dissimilar currents.'"[18] Hall's attention to the convoluted features "moments" contain is helpful. It suggests an opacity created by global circulations and "grounded normativities"; by concurrent affirmative and negative soundings. This is the "moment" that links Gyibaaw to world historical time; a time of global Indigenous unrest and anticolonial striving.[19] Taken together, Gyibaaw's combination of local and extralocal mediations and opacities cannot be reduced to Indigenous or non-Indigenous elements. Rather, their entanglement, and the elaboration of a sound both place-specific and globally resonant, becomes Gyibaaw's story of "fugitive Indigeneity," one that connects it to an Indigenous North Coast and, in another way, to black metal. To begin to unpack this claim, I need to say more about the musical sound in question here.

Indigenous Black Metal: A Brief Excursus

Gyibaaw was more than a story of music-mediated Indigeneity. It was also a contribution to the history of popular music and the centrality of Indigenous contributions to its expanding genre forms. By and large, the contributions of Indigenous peoples to black metal have been ignored. This is surprising, given the fact that for many critics writing since the Millennium, black metal stands as a propitious music case for interrogating the contemporary politics of authenticity, as well as the untimely return of global antimodernist movements.[20]

Black metal music is suffuse in its professions of authenticity and antimodernism. As Benjamin Noys explains, it is marked by a desire to cut through what Jean-Jacques Rousseau once called "the wound of reflection"—the compound effect of modernity's various disenchantments (lately made worse by neoliberalism).[21] For Timothy Morton, black metal's appeal further extends from its ability to link these yearnings to the psychosocial dislocations of ecological crisis.[22] Black metal, Morton explains in his characteristically florid prose, is music for the Anthropocene. This ecological thematic is important, but revealingly, Morton says little about the politics of race that also mark black metal. Nor do we hear of South American bands like Sarcófago and Mortem. All too often, the question of black metal operates without the question of Indigeneity. Its history moves in the classic diffusionist way: a scene first cultivated in Scandinavia in the late 1980s, then spreading across the musical subculture through elaborate tape trades and fanzine subscriptions. Eventually, the story goes, black metal finds its cultural appeal in Europe and North America after popular outlets begin to publish profiles of controversial bands like Emperor, Darkthrone, and Mayhem in the early 2000s.

But the diffusionist account of black metal overlooks key details. Obscure but extensively global circuits of Indigenous black metal have long existed alongside the European and North American scenes—and existed, in fact, with significant cross-pollination. The members of Gyibaaw were participant to some of these exchanges, as their music bears out (amalgamating, as it does, the chordal structures of Scandinavian bands, the rhythmic influences of South American bands, and the traditional drumming of the North Coast). "The supreme irony of the early 90s Norwegian black metal scene," notes historian Keith Kahn-Harris, "was that it was dependent on location within a country where devoting one's life to metal is easy due to the strength of state support (through education, social security etcetera)."[23] This stands in stark contrast to Jakarta, Quito, and Indigenous Los Angeles—where black metal scenes took shape without support from the state, and oftentimes in contexts of active state repression. What we find here is not simply an elaboration of a preexisting cultural trope, but a direct challenge to its assumed foundations.

Indigenous black metal is predicated on a dialectical reversal: whereas many of the most celebrated black metal bands were white middle-class responses to the spiritual ennui of postindustrialization, Indigenous black metal has arisen from contexts of ongoing industrialization and dispossession.[24] Gyibaaw is of a piece with this lineage. On the North Coast, as across the global resource periphery, the "Nietzschean aristocratism" of Scandinavian black metal falls away.[25] In its place, collectivist and anticolonial themes emerge—speaking to the extractivisms and dispossessions that have forced millions of Indigenous men and women off their lands. For hundreds of thousands of Indigenous people around the world, black metal music has come to constitute a propitious medium for enacting Indigeneity. It is an angry Indigeneity, as I explore below. The ethos is summarized by Yaotl Mictlan, a Utah-based band, rooted in pre-Colombian Maya: "Que hacer orgullosos de su pasado para enfrentar la lucha que viene!" (Remember your past for the battle to come).

Black metal mattered for the teenage cousins as they considered their Indigenous relationships to the North Coast. To Jeremy and Spencer, it mattered in a way that other negotiated cultural forms—like hip-hop or country music—evidently did not. This is surprising, given black metal's opacity. Black metal music "often sounds close to being a formless noise, but backs away from doing so at the last moment," as Kahn-Harris puts it.[26] It is characterized by blast-beat drumming, distortion, feedback, and searing indecipherable lyrics. Black metal guitarists use tremolo picking and extensive distortion, nondiatonic tritones and power chords (e.g., open fifths with the third omitted) rooted in different tonal vocabularies than most rock music. The intention is instability and intransigence, and it is complemented by a pageantry of dimly lit atmospheres, candles, and black hoods.[27] Across a range of expressive examples, black metal scenes operate as defiantly subcultural, antisocial, even hostile zones to outside interest.

There is much in Gyibaaw's music to suggest a kind of Indigenous Romanticism, an art seeking to restore the inward-facing sincerities of ritual and tradition at a time of territorial destruction and spiritual despair. Yet the story of Indigeneity is emphatically a story of retrieval *and* reinvention, as Gerald Vizenor has shown.[28] Indigenous black metal complicates black metal's Romanticist self-narrative. Today, there are hundreds of black metal bands around the world that self-identify as Indigenous.[29] A survey of this work suggests both a cultivated use of common tropes and their creative rearticulation too. Central here is an interest in combining European sounds with so-called traditional ones: drums, rattles, flutes, and singing styles that extend to traditional gutturals and language forms. Against postmodernism's blank parody, Indigenous black metal musicians often use these combinations to proclaim profound fealties to places and identities, translating black metal's ornately symbolic graphologies into detailed place narratives. The very tractability of traditional music is distorted, with canonical expectations questioned in turn. It is here that we again find the "fugitive aesthetics": opacity, a black metal strategy to "blur the parts together to create an atmospheric wash of sound."[30] In Gyibaaw, this strategy will sometimes press an opposing tendency, one which returns us to Hall's conjunctural moment. This is the ideal of a communicative musical Indigeneity: of music's professed power to claim fealty to a territory and to the refusal of its capture. In the punk-houses and metal venues where Gyibaaw played, few in the audience would have been privy to this message. The offering, rather, was a material and symbolic surplus, a sound that confounded audience interpretation while supporting powerful connections to Gitk̲'a'ata.

Early in my research, I found a remarkably lucid exposition of Indigenous black metal on an anonymous blog post. Featured on the defunct blog *Metal Nation*, "Understanding Black Metal; a Native American Point of View" considers some of the connections black metal can make available to Indigenous peoples:

> As a Shawnee Indian I believe our people's history, spirituality and suffering prepares us to understand Black metal better than most Americans. Black metal tends to open spiritual doors; being evil or good. Its ambience and brutality resonate in the regions that most cannot see. The regions that us Shawnee call the spirit world, or spirit place. I believe Black metal taps into this opaque and mysterious place. This hidden place, the region between the light and darkness is understood by the world's Indigenous cultures; being Scandinavian or American. Black metal is a lifestyle and a medium for spirituality in the truest sense of the word. The proof lies in the corpse paint championed by the Scandinavian Black metal bands. Our people donned the same basic style of paint as did many other Indigenous cultures. Black metal's grandfather is nature, its mother is spirituality, its father is brutality, and its children are a combination of all three.[31]

I emailed the quote to Spencer before we had arranged to meet at Tim Hortons one day in January 2014. I brought it up during our interview, and his expression immediately suggested it had resonated. But Spencer was also careful to specify that Gyibaaw's embrace of this sound was not immediate. "We actually started out as death metal," he noted. "And then that became blackened death metal. And then black metal, which we eventually began to call Tsimshian war metal." "It was never serious at first," Jeremy confirmed later. "We got into metal because of monsters and demons, the silly stuff you could do with all that. It took us time to find out what we wanted to do."

The turn to black metal was gradual. A variety of articulations were explored, until Spencer and Jeremy found the one that could "open spiritual doors." As their dedication to Gyibaaw grew, the band's initial humorous elements were modified. The name Gabow was changed to represent a more "authentically Ts'msyen" Gyibaaw (Jeremy). A darker sensibility began to preside over the songwriting. What black metal gave to a pair of seventeen-year-old Native kids growing up off-reserve, Spencer explained, was a vessel. "I was angry," he told me bluntly; "I would wake up in the morning, and right away . . ." We were now sitting in his truck, up a logging road near the Ecstall. Rain speckled the windshield; we were waiting for a break so Spencer could begin his hunt. "And I was learning about Duncan, and missionization," he continued, "and all the colonial violence perpetuated on our people. But then, I began to see something really big, and really meaningful, could happen in writing Gyibaaw."

For Spencer and Jeremy, participating in black metal meant a double negation: not only the structural exclusions of race in a white-dominated scene, but the decision to seek a community that celebrated marginality and abjection as virtues. It would be a mistake to read their embrace of black metal as an unproblematically affirmative decision. Black metal was, as the Shawnee blogger suggests, a music suffused with brutality. It was necessary, Spencer explained, that Gyibaaw sounded angry before it could sound good. "Under certain conditions," Glen Coulthard offers, "Indigenous peoples' anger and resentment can help prompt the very forms of self-affirmative practice that generate rehabilitated Indigenous subjectivities and decolonized forms of life."[32] For Spencer and Jeremy, music provided these conditions, and in ways that yielded unexpected results.

North Coast Stories

In April 2014, Jeremy took me on a drive around the industrial outskirts of Prince George, BC. A six-hour drive inland from Prince Rupert, past its rain-soaked mountains and overcast skies, Prince George feels like another world.[33] We passed old manufacturing buildings and empty parking lots. Jeremy pointed out spaces where secretive venues had once existed. Gyibaaw, he explained, played its first gig in a now-defunct space called the Roll-a-Dome. "We were always surrounded by a lot of shitty people there," he said with a laugh. "I knew a lot of the people who came to those shows. There were always folks you didn't want to know, and never wanted to see again." Spencer and Jeremy spent their teenage years in Burns Lake, a village a few hours north of Prince George. For them, along with Gyibaaw's other bandmates—Brandon Dyck and a Gitksan bassist, Norm McLean—Prince George was a big deal. For aspiring musicians in northern BC, it was "the place where you made your mark" (Spencer).

But Prince George was also a place where the brutality of the sound Gyibaaw pursued found its analogue in an everyday social reality. When sociologists Kim Anderson and Robert Innes reflect on the dismal policy support for Indigenous men's issues in British Columbia, they might as well be describing the city.[34] Across the province's far north, the broken edges of industry towns are for Indigenous peoples manifest instances of the repressive imaginary Fanon once described as a "world without spaciousness."[35] In Prince George, Jeremy told me later, you can get trapped. Nearly all exits lead to economic uncertainty, police patrols, or the complexities of the reserve.[36] As Tyler McCreary notes, the state's solution to the problem of surplus Indigenous men in BC is "the Aboriginal Worker," a subject produced to meet capitalism's resource extraction needs.[37] Jeremy knew many Native people who had gone to work for oil and gas. He knew others who had refused and fallen into drug use. This was one of the reasons, he told me, that putting up with a bad music scene was tolerable. His auntie, Cherill (Spencer's mom), had been so insistent that Jeremy and Spencer focus on music, she had gifted Gyibaaw the entire boot room of the 1980s trailer they shared ("Jeremy and I slept on bunk beds and jammed in the boot room," Spencer later added).[38] The cousins practiced relentlessly.

Music was the exit. "I used to put on my headphones, and just go out into the woods," Jeremy said. "I'd walk out there for hours in my leather jacket and army boots, just walking and listening to black metal." Walking is a common trope in black metal: the journeying act that enables "confrontation with the unknown."[39] Here, black metal's great Romantic exemplar is Caspar David Friedrich's (1818) *Der Wanderer über dem Nebelmeer*, the solitary male walker who contemplates an emotional landscape from on high. But Indigenous peoples have long recognized walking as a source of renewal with the land too. Walking the land is a way to enact one's identity—spiritual walks, cleansing walks, medicine walks. For Jeremy, walking old logging roads, listening to black metal music, began as a coping strategy. Eventually, it became a source for his music: the aliveness to the lands and the spirits that Gyibaaw solicited.

"Indigenous peoples," writes Kim TallBear, "have never forgotten that nonhumans are agential beings engaged in social relations that profoundly shape human lives."[40] For Ts'msyen as for other coastal Indigenous peoples, the forests, mountains, and rivers of the North Coast are crisscrossed with stories, beings, and living forces. Among these spaces, Ts'msyen artists can establish dynamic webs of relationships—expressive of what Gary Tomlinson calls "a grammar of metonymy."[41] These webs connect places and actions in ways that have no analog with Western discourses of music and landscape. The Ts'msyen anthropologies of Jay Miller, Susan Marsden, and Christopher Roth affirm the idea, but so do Jeremy and Spencer. As I gained some sense of these landscapes over my own trips up and down the Coast—from Cetacea Lab to Prince Rupert to Port Simpson to Kitwanga—I began to feel more comfortable asking them about the territorialities in their sound, and the places where the band played their music.

Here's a story I have been permitted to share. Along Highway 16 between Terrace and Hazelton, partially obscured from view by trees and a Canadian National Heritage plaque, stands Gitwangak, an old village site near the Kitwanga River. It is known colloquially as Battle Hill, as its steep ascent and conical peak made it a site of prodigious warfare in the late 1700s and early 1800s between Gitksan and other nations. Sometime in the winter of 2008, a small musical gathering formed on Battle Hill. It was led by Jeremy and Spencer. It featured several Gitksan: Gyibaaw bassist Norm McLean, Norm's father (Norm Sr.), and his brother, Robin. Under moonlight and a frozen sky, they began a musical dialogue. "We brought drums and flutes and did ceremony," Jeremy explained. "Everything was live—with all the sounds and spirits of the place mixed in." A recording eventually made its way onto Gyibaaw's debut, *Ancestral War Hymns* (2009), as the bonus track.[42] A succession of drums, rattles, and muffled voices—indistinct and limited by wind—it nevertheless evokes the sense of communion Jeremy described to me years later.

Bonus tracks are meant to be hidden, so it is not surprising that I was unaware of this one until Jeremy brought it up. A year and a half later, I was sitting at a table with Spencer in the visitor house in Hartley Bay. I was in the village to do work on the acoustics baseline, and the bonus track came up again. I was playing Spencer some of the recordings, noting the times and locations. The hissy sounds we had been gathering reminded him, he said, of the ambition Gyibaaw had long nurtured. "It was our dream to go to all these ceremonial sites in our territory and record the sounds, just like we did there, and have those sounds—the old fishing sites, the winds, the waves our ancestors paddled in their *wiixsoo* [war canoes]—playing through every song on our next Gyibaaw album. We wanted to overlap that entire recording with the album songs. Because the voice of our land is so integral to the music, they should exist together. We always felt the album was incomplete without that." I played him more recordings. Spencer said he liked the staticky and hissy ones the most: the ones overdriven by storms, heavy rain, and crashing waves. They were similar, in this way, to the Battle Hill recording. Listening to Spencer that evening, I remember wondering if these distorted qualities contained with them a different source of value than I had access to. Perhaps they suggested sounds that would be kept underground, away from mainstream listeners and mainstream logics. Or maybe Spencer was telling me something about the ethics of recording, as he understood it. Not unlike Hermann and Janie, he held a different relation to the noise of this place. In its opacity, perhaps he could hear the voice of Gyibaaw, or some indication of its presence, reaching audible form, infusing the music around it.

Ancestral War Hymns is the only full-length album Gyibaaw released. It is a sprawling work of forty-seven minutes. On its cover is a solitary warrior in a war canoe adorned at the corners with eagle claws. On the inside is a small grainy photo of Gitga'at Territory, in black and white. I came to know Gyibaaw's music through various mediated forms: Myspace pages, shaky YouTube videos of live shows, but *Ancestral War Hymns* most of all. Jeremy and Spencer would each lead me through it, pointing out different themes and details as we listened together. "Gyitwaalkt," the opening track, is a Sm'algyax term roughly comparable to *warrior*. "We wrote this song to portray the ceremonial fast of a warrior and the tapping into a spirit world," Spencer said, "the act before they go into war." We are huddled in my small attic room in Prince Rupert, drinking tea. Spencer has paused the track to note how the intensification of the kick drum (which occurs around the one-minute mark) enacts a transition into sea-based battle: "That comes in when you are actually seeing the enemy," he continues, "when you can imagine the heart is just racing and pounding." This transfiguration is a typical Gyibaaw strategy—when the song is played live, it effects transformations into ancient warriors, ocean travelers, and more. Another example comes in track two: "Ueesoo," Sm'algyax for *canoe*. "The first thing you would see in an upcoming battle is the war canoe," Spencer explains. "Music was always a part of warfare, and that's what inspired us to go to the canoe." With "Ueesoo," the material movements of the ocean are rendered in terms of pulse, movement, and rolling 4/4 rhythms. The Locrian mode cherished by black metal supplies an expressive syntax—with a dominant chord affixed to chords modulating underneath to create layers and wavy undulations. The result is a chaotic battle scenography—crashing waves in cymbal splashes and spears in guitar lines that dive down from the sky.

For Spencer, Gyibaaw became an opportunity to develop his interests in traditional Ts'msyen drumming. "We wanted traditional drums and beats to be the backbone and the heartbeat of everything," he explained, "because traditional drums are the heartbeat of our territory." Another notable aspect of the music, present in the drums, is its concern with temporality. *Ancestral War Hymns* features frequent sonic appeals to "intergenerational time," Kyle Whyte's term for movements "embedded in a spiraling temporality."[43] Whereas most tracks, like "Transcending into the Spirit of Gyibaaw," deploy a range of time signatures, suggesting an interest in temporal disjuncture, others, such as "Winter Emissary" and "Gisigwelgelk," move deliberately out of joint. They are plodding and obscure. They feature long flute passages and reverb-laden acoustic guitar that combine with the tempo to suggest a kind of intermundane portal—a semantic figure anthropologists have long associated with Ts'msyen culture.[44]

A third notable feature of *Ancestral War Hymns*, and perhaps its most arresting sonically, is Jeremy's voice. Voice, Jacques Lacan famously observed, designates a capacity to extend the organism "to its true limit, which goes further than the body's limit."[45] On *Ancestral War Hymns*, voice is Gyibaaw's principal site of transfiguration. Growls become screams, animal voices become human voices, shrieks morph into icy pellets of rain. These moments showcase Jeremy's ability to manipulate his voice's glottal capabilities. His voice announces Gyibaaw and sometimes becomes Gyibaaw, mediating *Naxnox* through rasping breath, illuminating the spaces around. A potent instance is the ten-and-a-half-minute "Nekt," which commences with a sampled roar of a grizzly bear, becoming a high-arching howl as Jeremy's voice becomes the movements of the legendary Gitksan warrior.[46]

In one of our interviews, Jeremy suggested "Nekt" had a contemporary meaning. It was a kind of preparation, he noted, for getting Indigenous peoples to confront the Enbridge Northern Gateway pipeline. "We needed creative ways to get language and culture for our young people to connect," he told me in a May 2015 conversation. "We saw our role as musicians, and we used that to express what we needed to about the need for Indigenous warriors in that moment." Similarly, Spencer would describe the track "Itlee Tsimshian [I Am of Tsimshian Blood]" as Gyibaaw's response to the present-day colonial occupation of the North Coast. He also said it was one of Gyibaaw's most serious songs. The move from playful to serious was a characteristic feature of my time with Spencer. He is one of his nation's political leaders, with a big Longhouse voice and a bright warm laugh. But with "Itlee Tsimshian," Spencer also spoke to some of our communicative limits. Its ultimate meaning, he explained to me, was exclusive, directed to listeners belonging to his nation and, perhaps, to the members of Gyibaaw itself.

The distribution of *Ancestral War Hymns* was handled by a legendary black metal label, Ross Bay Cult. Centered on the influential Burnaby, BC–based black metal band Blasphemy, Ross Bay is a coterie of musicians and promoters who first gained notoriety in the early 1990s for conducting grave interments at the upscale Ross Bay Cemetery in Victoria, BC. The cemetery has a strong colonial legacy. It is lək̲ʷəŋən (Lekwungen) peoples land. It is also the resting place of many of British Columbia's most notorious colonial figures, include its first governor, James Douglas—namesake of the Douglas Channel that cuts through Gitga'at Territory and is now its proposed shipping route. Whenever they were passing through the area, Gyibaaw made trips to Ross Bay. They would wait until it was dark and then combine black metal libations with Indigenous smudging ceremonies. Friends of the band described these evenings to me: curiously intense amalgamations, playful and serious, sacred and profane, not unlike the music itself.

The Ross Bay distribution label helped to launch Gyibaaw into the global circulations of black metal culture. Here, they were also supported by sociotechnical developments. The early 2000s was the acceleration stage in the digitalization of music culture. Soon after its release in 2009, discussion of *Ancestral War Hymns* developed on MySpace accounts and online metal blogs. The interest supported Gyibaaw's involvement in the legendary tape trades that have long supported dialogues between Scandinavian and South American black metal scenes. Some of these exchanges helped to set the stage for Gyibaaw's invitation, from a Colombian tour manager, to tour South America. At this point, Jeremy and Spencer were barely past their teens. Spencer's mother and Jeremy's auntie, Cherill Pahl, insisted she tour with them as a bodyguard. For several months, Gyibaaw visited ranching districts, mountain villages, and venues in Bolivia, Ecuador, and Colombia. "I remember the crowds," Spencer recalled. "In Ecuador it was these Indigenous farming communities who came out. We stayed at this barn in this big bunkhouse. I remember we were playing soccer with all the kids. These six-year-olds with Destroyer 666 shirts on. And then we played this show with two local bands and like five hundred people showed up. It was just nuts, the way they were dancing and cheering us on. These big communal dances. It was incredible."

By the time they returned to Prince George, Gyibaaw's reputation had grown. The band exuded a new confidence. "Our lyrical content is an expression of our culture and spirituality," the band announced in a 2010 tour statement: "We live it, and we acknowledge the fact that it goes back thousands of years. The lyrics are about 60–75% in 'Sm'algyax,' which is the language of the Tsimshian people. By using our ancient language, MUCH more feeling can be expressed about our homeland, the ancient spirits, our warrior culture, the natural world around us, and all creatures who dwell on this Earth."[47] In 2011, Gyibaaw played a rapturous set opening for the US black metal band Wolves in the Throne Room. A review in Vancouver's *Georgia Strait* remarked on the band's decision to commence their show with a land acknowledgment to the Musqueam, Squamish, and Tsleil-Waututh hosts (a completely unusual gesture for popular musicians at the time).[48] Jeremy and Spencer brought animal skulls onstage and adorned the boom stand with rattles. When I reached out to Nathan Weaver, drummer for Wolves in the Throne Room, he recalled his incredulity. "So often with bands, you basically know sound, their friends, their story," he noted. "With these guys, I remember being completely unable to place their sound. It came from somewhere else . . . but where was that?"

By this point, bands were increasingly looking to have Gyibaaw tour with them. At the same time, Jeremy and Spencer were spending more and more time in Hartley Bay. Old enough to travel to the village on their own, the cousins were also participating in a range of cultural and harvesting activities. In the summer of 2010, they represented Hartley Bay in the annual Gathering Strength canoe journeys that convene Indigenous youths from up and down the coast. "The hair!" Hartley Bay resident Mary Danes laughed when I asked her for recollections. "All these teenagers paddling with normal clothes on and then those two with their leather jackets and long hair." Mary hadn't listened to Gyibaaw when I interviewed her in January 2015. But she understood that the cousins had a project that was requiring that they spend more and more time in their territory, with family and community.

As Gyibaaw's success grew, so too did their willingness to question black metal's problematic elements. During one long conversation, Spencer shared some of the frustrations Gyibaaw had helped him formulate. There was the sexism and the drugs, but also the scene's lack of commitment to the authenticity it projected. It came up in the way Gyibaaw's audiences had appraised their purported icon: the wolflike figure of Gyibaaw. "If the mainstream Bible didn't demonize things like wolves, black metal kids wouldn't have thought twice about them," he explained. "To the black metal scene, it was about playing dress-up and trying to be scary in a Christian context, while we had a culture grounded in millennia of teachings around these beings."

Black metal's wolf obsessions run deep: from the adopted name of Mayhem's Varg Vikernes (*varg* being Norwegian for *wolf*), to Black Funeral's "Der

Werewolf," to Pest's *Hail the Black Metal Wolves of Belial* (2003), black metal bands exalt wolves as embodiments of spiritual inclinations, mythological curiosities, and shadowy aggressions. Black metal bands celebrate wolves to defy appraisals of wolves in Western culture: as villains, vermin, or worse. In his frustration, Spencer seemed to implicitly grasp Stephanie Rutherford's remark: in North America, both the venerating and the defiling of wolves can serve the same desacralizing settler colonialism.[49] He was finding it increasingly necessary to challenge the subculture claiming to oppose the mainstream through its lupine glorification. Wolves impart different relationships in Ts'msyen cosmology, he explained, with deep histories in the North Coast.[50] They are denizens of unique societies, with powers of solicitation and cunning that transcend the possibilities of attachment via identity alone.[51] Wolves are to be acknowledged, he added, but not symbolized. "And the Gyibaaw is a wolf," I added back, trying to link the discussion to the band. Spencer corrected me: "Not always. The Gyibaaw is a wolf, but in some contexts, it is a Naxnox."

During one of my stays at Cetacea Lab, I recorded a spectacular sequence of wolf howls. As the evening calls rang out across Taylor Bight, they were joined unexpectedly by the sounds of killer whales, emitting tonal blows nearby. The animals offered a chorus I've never encountered since. I stood with a group of other interns on the balcony, totally transfixed. Jeremy loved this recording story, asking me to recount it to mutual friends years later. It was a story of transformation, he suggested, not unlike that of Otter Woman—a story Norm McLean's brother Robin had recounted to him, about a shapeshifter along the Skeena; a girl who had politely asked for a ride home, and then sat in the rear seat covered in darkness, concealing her shapeshifting face from view. Life on the coast unfolds in strange rhythms and surprising transformations, these stories suggest. Danger can lurk in unexpected places. Wolf howls can be the voice of the land, but they can also be prophecies of changing relations to the land—heralds of dark forces. "Perhaps," writes Carla Freccero, "the persistence of the becoming wolf of humans in the popular imagination has something to tell us about the survival of humans and wolves on a damaged planet in apocalyptic times."[52]

In late 2011, Gyibaaw started using a former skinhead named Daniel Gallant as their tour bus driver. Gallant's involvement reflected the band's growing concern with the new elements visible at their shows. Over the phone with me in 2014, Gallant described his familiarity with the hate groups that have long operated on the fringes of Prince George's metal scene. They were coming to black metal shows, he explained, because the noisy spectacles were good recruitment spaces. A February 9, 2011, article in the *Prince George Citizen* confirmed that "Nazi flags, white power tattoos, racially adorned clothing, and racist online chatter . . . are familiar sights in Prince George." Wanting to learn more, I reached out to Lee van der Kamp, a close friend of Jeremy and Spencer's. "Nazis were fascinated with Gyibaaw," he recalled. "They would come up to Jeremy after shows and praise him. There was always a weird disconnect because they said they were in awe of Gyibaaw, but they didn't want to know too much about the actual musicians." There were other stories: Bassist Norm McLean mentioned the increase in white shoelaces at shows in northern Alberta—a telltale sign of Nazi filiations. Spencer was working the merchandise table after a show in Montreal when he was approached by a gaunt figure with a swastika belt buckle. "He was like, 'I love your message and the stuff you are doing and the imagery of your music.'" Spencer continued: "He created these wild connections between our message and right-wing black metal. It was a really crazy experience."

Articulations can be dangerous, Stuart Hall once observed, because "the grooves" instilled by previous articulations can be hard to undo.[53] Hall's record player metaphor is instructive. As the Gyibaaw band members were increasingly aware, their music was supporting a range of meaningful connections—some affirmative, some oppositional, and some that the band was actively seeking to resist. In spring 2011, Gyibaaw started touring with Inquisition, a well-known black metal band from the United States. During a drive between venues, the group went swimming in a lake. When Daniel Gallant took off his shirt to jump in, Jeremy and Spencer were shocked to discover Inquisition members praising the ex-skinhead's swastika tattoos.[54] They would later learn about a split-album LP that one Inquisition member had recorded with Deathkey, a band with neo-Nazi sympathies. Gyibaaw's experiences with the extremist subculture were increasingly being covered by fanzines and blogs, debated, gossiped about, turned into a spectacle. It was, for both cousins, a breaking point, as sudden as the band's ascent. In mid-2012, on the cusp of international fame, Jeremy and Spencer stopped playing Gyibaaw.

Conclusion

I am sitting in Johnny Pahl's living room in Prince Rupert. Around me are wood-paneled walls laden with Ts'msyen crests, country music playbills, and old family photos. There's an oil painting of the younger Johnny Pahl with his three brothers—all wearing bolo ties. "JP" used to ride the Canadian National Railway as an "Indian cowboy," Jeremy tells me, playing Hank

Williams and George Jones. Jeremy and Spencer's latest project is playing on the speakers: the Saltwater Brothers, the old-time country music band they formed soon after Gyibaaw ended. Several times during our conversation, Johnny will come in to listen, tapping his hand to the beat in an expressive way that makes Spencer laugh. Later, he will join us and recount all the places in the North Coast where he used to play: Prince Rupert, Hazelton, Sunnyside, Kemano, Hartley Bay, up and down the Nass. From Jeb Nelson's brass band (1900s), to the Hartley Bay Five (1960s–70s), to the Hartley Bay Choir (1970s–present), to the Saltwater Brothers (2012–), to still newer sounds, Ts'msyen peoples continue to produce innovative music. This expanding sonic archive is a correlate to the "moditional" economy (modern + traditional) historian John Lutz discusses in *Makuk*, his Ts'msyen labor history. In this chapter I have tried to consider the political dimensions shaping one "moment" of this history. I have considered how layered forms of opacity in Gyibaaw—intimate and place-specific, but also genre-specific and world historical—serve to explain both the aesthetic strategy of the music and the different ways it would be received. Through their fusion of "contradictory forces," Gyibaaw's "practices of incomprehensibility" made for a visionary music that drew fans from around the world. It changed the lives of its authors and restoried the landscapes of an Indigenous North Coast.

In 2008, Indigenous black metal was rarely invoked beyond the subcurrents of a subcultural black metal. Ten years later, there would be government-sponsored arts grants for a variety of experimental Indigenous art forms, and Spotify playlists hyping Indigenous metal acts—sonic capital for a global art-finance nexus. *Ancestral War Hymns* would become a collector's item, pursued on music subreddits and hipster blogs. The articulatory conditions are decidedly different today than when Gyibaaw first released the album. In response to Idle No More and interrelated resurgence politics, Canada's government has enacted a broad suite of reconciliation strategies: arts funding grants, land acknowledgments, institutional positions. Indigenous musicians continue to confront its widening accommodative structure, one that motivates Dylan Robinson to warn of reconciliation's "affective aura."[55] Meanwhile, the subcultural currents of music continue to change. New projects, like Blackbraid or Brutal Morticínio, are elaborating new forms of Indigenous black metal. If some build on Gyibaaw's approach, others engage queer, feminist, and Afro-pessimist perspectives and push beyond it.[56] Music is affirmed as a dynamic process, perhaps the most dynamic of all cultural forms. Music flows like the Skeena, a river fed by many tributaries, in rhythms and movements unique to music itself.

And so we should leave this story with a different sense of place: with an appreciation of the "active sense of presence" some listeners find in music, and with an appreciation of the Gitk̲'a'ata connections to Gitk̲'a'ata. Gyibaaw was a multiagential music that affirmed the creativities of an Indigenous North Coast, moving below and occasionally alongside conventional frames of politics and aesthetics. Amid the torsions of global culture and industrial development, it was a story of dangerous music and dangerous articulations (some all too familiar to Indigenous peoples). This helps to suggest, once more, why the aesthetics of Indigeneity routinely turns to opacity and the desire to protect sounds from the possessiveness of others. Nevertheless, it is possible to look back and observe in Gyibaaw a band that left deep traces within the archives of the North Coast. Jeremy's favorite story about the band begins after a show at the Nisga'a Hall in 2011, when an unnamed father approached him with his twelve-year-old and said, "Thank you for making my son proud to be Indigenous."

5

SMARTEST COAST IN THE WORLD?

DIGITAL SOUND AND ENCLOSURE

The Hartley Bay ferry dock is one of the great meeting places of the North Coast. Although it can sit quiet at times, it will at others uniquely exude the wonderful diversity of lifeways that marks the region. Walking along the boardwalks on summer afternoons, I would encounter crews of scientists disembarking from government boats, construction workers from eastern Europe and Mexico, whale research interns from Scotland, and American yachters making pit stops en route to Alaska. But the ferry dock is above all a platform for Gitk'a'ata. The rectangular wood plinths are where teenagers chop fish, elders gossip, fishermen refuel, and Gitga'at Guardians reunite with their families after a day out on the water. At approximately 12 p.m., two days a week, the *Tsimshian Storm* pulls in from Prince Rupert, and the residents of Hartley Bay emerge from their houses, hop into golf carts and Humvees, and drive down the gangplank. The dock becomes a bustle of voices, laughter, and the hectic and often exuberant loading and unloading of boxes.

As Tarleton Gillespie enjoins, "all platforms moderate."[1] Platforms shape possibilities for connection. They encourage some exchanges and delimit and exclude others. They can be the basis for face-to-face dialogues, mass cultural displays, or the last cellular transmission before the signal is gone. In enabling such possibilities, Hartley Bay's platform provided me with something else: an opportunity to consider just how unique life on the North Coast can be. As such, it provided me a chance to observe sociality, nothing less than a "grammar of . . . relational imagination, action and potential," in Max Haiven's words. Sociality is about the ways people relate to one another—culturally, ecologically, historically.[2] And it is because of sociality that my encounter with Ocean Networks Canada (ONC) felt so propitious. Coming down to the Hartley Bay ferry dock one spring day in 2014, I found a small group of people chatting with Janie Wray. Janie introduced me to one of them. Tom Dakin was a former Canadian naval intelligence officer. He had recently joined ONC as a sensor technologies development officer. Tom had heard about my eco-acoustics baseline work with the community, and he was eager to trade notes as we sat together on the ferry ride up to Prince Rupert. Unbeknownst to me at the time, ONC had already been making inroads with Hartley Bay, as it had with several other coastal communities. In a few short years, ONC's Smart Oceans project would become the central expression of a new era of environmental governance on the North Coast. The initiative would become a topic of discussion for many concerned about the risks of development too, and specifically as the best way to respond to and manage them.

Compared with the Hartley Bay ferry dock, Smart Oceans entails a very different kind of platform. It is given in the name itself: Smart Ocean, a spatially distributed infrastructure generating big data in the form of sensor measurements, imagery, and recordings.[3] "By continuously capturing, archiving, and delivering data from the ocean," an ONC report explains, "these observations support scientific study and inform decisions about earthquakes and tsunamis, climate change, coastal management, conservation, and marine safety for the benefit of Canadians."[4] As Orit Halpern and Robert Mitchell cogently note, the discourse of smartness projects "an imaginary of crisis that is to be managed through a massive increase in sensing devices."[5]

As a technological system, smartness requires the continual mediation of new network nodes and sensor data—cost-intensive affairs, but necessary for providing the best possible response to new risks. On the North Coast, a Smart Ocean would be a timely proposition. Its promise of efficient and transparent information comes at a time when environmental challenges have plainly exceeded traditional state capacities, with ocean spaces increasingly beset by multiple ecological uncertainties. Smart Oceans would be just one of the new sensing infrastructures I observed taking shape in the region.[6] In exchange for the right to record and produce data, ONC's initiative promised access and integration into powerful systems of environmental knowledge. Through its data platform, Oceans 2.0 (redubbed Oceans 3.0 in 2022), many residents would, indeed, join a project that promised to extend their "spatial and temporal reach," including via "real-time sound-in-the-sea data in critical coastal areas."[7]

My concern in this chapter is to consider the changing perceptions of coastal life being instigated by Smart Oceans. Across a series of sites on the North Coast, the digital sounds recorded by this project reveal a particular trajectory, or motive thrust, expressive of broader development ambitions. In Smart governance projects around the world, we are witnessing the rise of networked forms of sociality that privilege sound primarily in terms of informational functionality and signal. Tone and the polysemous gesture withdraw, and living senses become increasingly valued as elements in concatenations of data flows.[8] A kind of sensory de-skilling is evinced here in as much as listening acts are replaced by digital operations with a resulting diminution of their value. This is all, I will argue in this chapter, an effect of the enclosure processes ongoing in myriad marine spaces today.[9] Smart Oceans enacts a "transformation of environment into a media surround."[10] In so doing, it supports the seemingly ceaseless penetration of digital sensing into the institutions and infrastructural spaces of daily life perception—from high school classrooms, to seafaring, to collective forms of entertainment. More than a sensing project, Smart Oceans is a product of state efforts to promote digital infrastructure technologies as the basis of its marine development policy. To support the goal of a predictable coast—one that can be made secure for shipping, while safeguarding its populations from the unknowable shocks of climate change—Smart Oceans will require a variety of social adjustments. While some of these adjustments appear small—as small as the hardware space required for a digital sound file—they can equally be profound in implication.

Smart Oceans was just gaining traction as my fieldwork on the North Coast was drawing toward a close. My remarks in this chapter thus lean toward the speculative and combine insinuation with exemplification. I also want to maintain a certain fidelity to the nature of my ethnographic encounter. This speculative vantage can help lend a sense of the attitudinal shifts that were pervasive across the spaces of my research. As Álvaro Sevilla-Buitrago notes, the importance of social normalization in enactments of enclosure remains overlooked. What sorts of incitements, collective behaviors, and mispercep-tions render this process into something ordinary and unremarkable? How might they feature in one of enclosure's most contemporary of modalities, so-called digital enclosure, whereby collective knowledges and resources are alienated from the social commons via digital technologies?[11] The rise of the Smart Ocean, I argue, is a story of enclosure, and also of enclosure's normalization in digitally networked form. By remediating socio-ecological relationships through its digital platform, Smart Oceans draws multifari-ous elements of life on the North Coast into new circuits of measurement, optimization, and control.

Mark Andrejevic reminds us that today's digital enclosures are not entirely novel. Rather, they depend on the continued material and social impress of "primitive accumulation"—the so-called initial enclosures that shaped the force of capitalism and colonialism.[12] A fuller examination of enclosure on the North Coast would have to involve a number of attendant processes—such as the continued arrogation of Indigenous fishing rights or neoliberal-ism's decades-long commodification of civil society.[13] My goal in this chapter is more limited. Focusing on normalization, I aim to canvass the various and oftentimes subtle social adjustments a new mode of governance would in-troduce. The assemblage form—with its attention to the mediating forces of environmental spaces, sensors, humans, and datasets—once again provides rich vantage for making sense of the plural agencies involved in this process. While Smart Oceans suggests various moments in which affirmative, unas-suming acts of "assembling enclosure" can occur, this chapter emphasizes the sonic dimension.[14] To a significant degree, we can observe enclosure in the form of an artifact with a surprising utility to governance on the coast: digital sound.

I begin the next section by considering one early instantiation of the Smart Ocean: its so-called Community Observatories. After charting mo-ments of social normalization here and across other sites where ONC would introduce its Smart Oceans project—including high schools, job training programs, and community councils—I turn to enactments of digital sound. Attention to this object, and the social relations it composes, reveals significant changes to sociality on the North Coast. This idea is developed with respect to one particular expression: the spectral density probability map. It is here, I find, that analysis can perceive broader commitments to a predictable, if never fully knowable, North Coast.

Community Observatories

In the fall of 2014, ONC approached the Dodge Cove community with a proposal to "install a 10′ × 10′ shore station" just offshore from Digby Island.[15] The station was to provide "real time monitoring" through a hydrophone and other sensors capable of measuring marine environmental changes.[16] It belonged to a novel governance idea ONC was introducing to several other communities—the so-called Community Observatories project.[17] The first person I spoke to about it was Des Nobels. He was, to my surprise, supportive. "With Nexxen proposing what they are proposing," he explained, "we need this monitoring." In response, I pointed out that an ONC brief had explicitly mentioned how the hydrophone data from Community Observatories could be used to "help . . . industry access local data to support operations."[18] Des was not surprised at this. He was quite aware, he said, of ONC's connections with oil and gas—in fact, it was Des who later notified me of a May 2015 article in the *Globe and Mail* explicitly reporting on this issue.[19] But Des was insistent that the community's data needs were not being satisfied by government. What was on offer was ONC's Community Observatories.[20] For Des, the potential benefits of these systems—the real time monitoring, the ability to share data with scientists—outweighed the risks of giving data to private interests who might already have that data anyway.

Des was hardly alone in his assessment. Many people I spoke to in Dodge Cove and Prince Rupert expressed support for Community Observatories. And other manifestations of them would, indeed, begin to appear in other places: Ridley Island, Kitimat, and farther south along the coast. One reason for their popularity, it seemed to me at the time, was a kind of "early adopter" ideology: people saw Facebook posts and community chats about ONC and wanted to be ahead of the curve. Being associated with the observatories "seems appealing if others on the Coast are involved too," noted Chris Picard, the Gitga'at Nation marine planner with whom I had developed the Gitga'at Acoustic Baseline. ONC would begin its engagement process with the Gitga'at around this time (2014), establishing a Community Observatories system in the territory several years later. Like the others, the Hartley Bay iteration was customized in consultation with local leaders and offered localized opportunities for year-round ocean monitoring.

Besides physical connections to Smart Oceans, these collaborations offered something else deemed important, a social process increasingly present in other aspects of coastal life too: the digital network. More than a model of organization, digital networks are, observes Wendy Chun, powerful ideologies of community in action. They fuse "imagined synchronous mass actions" to create an imagined togetherness—"a 'we' that moves together through time."[21] This idea suggests how support for Community Observatories in Dodge Cove, like the support I found in other communities, might relate to a perceived absence, namely, the government's. Des held long-simmering resentment toward the province for its mishandling of local fisheries. Down in Hartley Bay, Chris Picard favorably compared the patience of the ONC scientists to the "turnstile approach" of government-led community engagement. Both pointed to an important irony: Smart Oceans' appeal was premised, at least in part, on a distrust for the state that was funding it.[22] But a further ideological element lay in the fact that this distrust was not a unifying force. "A key premise of smartness," Orit Halpern and Robert Mitchell note, "is that while each member of a population is unique, the population is . . . limited in its perception . . . that smartness emerges as a property of the population as a whole only when these limited perspectives are linked via environment-like infrastructures."[23]

Digital networks decompose structures before re-creating them in the networked image. "The network means to address matters of scale," Stefan Helmreich observes, "to connect the individual to the aggregate."[24] Communities sought integration into the Smart Ocean, it seemed, because the network suggested a larger totality through which they would be empowered. State regulators had positioned Dodge Cove as an operational endpoint they could manage from above when they wanted to, and then ignore when they didn't. With its emphasis on data sharing and participation, the Community Observatories scheme suggested lateral and continual relationships among partners. It was not often that I heard Des agree with Jason Scherr, the environmental sustainability manager of the Port of Prince Rupert. But both had become convinced of the value of ONC's project. By 2017, three Community Observatories were operational in the region: at Kitimat, Ridley Island, and Dodge Cove. Another was taking shape in Hartley Bay.[25] Partnerships with noncommunity groups had also appeared by this time, including with the City of Prince Rupert, the Department of Fisheries and Oceans (DFO), and the Prince Rupert Port Authority.

In this expanding arrangement, I began to see how Smart Oceans' production of the network as community could offer dividends. Many of the people I spoke with saw potential in the robust and freely available user data Community Observatories promised. Even where specific applications remained to be worked out, they seemed to concur with the claim made by one ONC brief, which stated that "accumulating time-series, benchmark data will foster confidence amongst coastal communities as they engage first-hand in protecting and sustaining the economic, environmental, social, and cultural health of our ocean and coasts."[26] The title of the brief, "Product and Services Deliverables Report—Transport Canada," is revealing. What is not communicated as readily to the partners are the limited terms of access Smart Oceans entails. Once recorded, Smart Oceans data is not owned by the community, with a few exceptions. Rather, it becomes accessible to them in the form of streams and downloads. Sampling rates can be adjusted and source points moved around, but property rights, patents, and Memorandums of Understanding (MOUs) are less pliable. With continuous monitoring, these too must be normalized as accepted tools of marine record keeping.

Digital Marine Education

In 2017, North Coast Community College (later rebranded Coast Mountain College) began to offer a community-based observatory education program. A central stated goal of this course was to acquaint students with the marine observation technologies necessary to marine governance in the North Coast. Tellingly, the approach taken in the course emphasized the sociality I have been speaking of: "a network of students and teachers involved with the community-based observatory project."[27] Participating students don't simply learn how sensors capture data; they learn how to themselves become participants in the community capable of integrating their own sensings with platform-based observatories. The college unveiled another course in 2017, too: Instrument Technology, designed to "introduce students to marine sensor technology, with an emphasis on underwater cabled observatories and shore-based coastal weather stations." Both courses were sponsored and codeveloped by ONC. They were testaments to new kinds of pedagogy, another moment in the normalizations and enclosures of a smart ocean.[28]

Not unlike my experience with Des and the Community Observatories, I was surprised by the grim pragmatism that seemed to characterize local support for ONC's learning initiatives. The clearest example was Ken Shaw. I had first met him when I was researching noise at the Fairview Terminal (chapter 3). He was a gardener who taught agroecology courses at the college. He was also an astute critic of smart meters and a Noam Chomsky fan. When I went to his office to ask him about the ONC courses, a knowing smile appeared across Ken's boyish face. "Well, they get trained how to use the data infrastructure from their smartphones," he said. "So, these are things that they complete and upload quickly. The students like knowing they can be involved in doing something right away." Ken added that student requests for "immersive learning" had been a factor in the college's decision to feature the courses. I responded that the only immersion in question seemed to be inside the smartphone screens students already had access to. But Ken replied that cell phones and tablets had become on-the-job training for many forms of applied environmental research. The college, meanwhile, could no longer afford to generate expansive field-training courses with the resources it had. In an era marked by funding cutbacks and digital tools promising new cost optimizations, students' ability to provision their own material needs was both institutionally necessary and perhaps even an employable skill. It was not an ideal outcome, Ken reflected, but it made sense.

Canada has become a leading international exponent of ONC's pedagogic model in recent years, combining commitments to "Sustainable Marine Development" with an "Oceans Supercluster" that seeks to galvanize job opportunities for technology builders, data analysts, and IT.[29] "Ocean literacy," as it is often called, is premised on the idea that maritime awareness can be improved by using sensor data and that, through its real-time flows, "students can experience the ocean from classroom locations."[30] Materially, ocean literacy is supported by the same things that constitute the smart ocean, including undersea cables, broadband internet, radar, Automated Identification System (AIS) ship tracking technology, and hydrophones positioned in "high traffic locations."[31]

By the time I left the coast in 2018, ocean literacy had become an established pedagogy in numerous local secondary and postsecondary institutions. It appeared in various guises, including Ocean Education (sponsored by the National Geographic Society), Turn Down the Volume! (sponsored by DFO, or Fisheries and Oceans Canada), Ocean School (Google Classroom), and Great Bear Seas (the Great Bear Rainforest Initiative). Ocean Networks Canada has been a particularly eager exponent of ocean literacy. Along with my broad observations of classes featuring ocean literacy content at the Northwest Community College, I came across a suite of related policy brochures and briefs. One instance is Ocean Sense—ONC's high school lesson plan that uses "phenomenal" data platforms to allow users to "dive into the ocean" in the medium of visual and sonic data.[32] Like the college courses, Ocean Sense seeks to familiarize its users to the prospect of professions geared to long-term data collection. Its web content includes "specific community-based observatory data pages" where users can access the latest hydrophone recordings, as well as opportunities to "virtually join" on deep-sea research expeditions.[33]

More than new lesson plans, these courses pose shifts to the assumed function of education. In the process, the North Coast becomes reconceived as an ecological space knowable through instrumental models of inquiry-based learning and the microtasks of signature identification from digital recording. Like its superannuating Smart Ocean, Ocean Sense normalizes modes of learning premised on data-based sensing. It deflects attention from linear constructs of narrative and causality, instead privileging efficiencies of comparison, extrapolation, and pattern recognition. In so doing, it echoes the networked governance vision of Smart Oceans. In the *Ocean Sense Teacher's Manual*, community-specific data pages are further proclaimed as a "one stop shop" that provides "easy navigation to the data and intuitive data analysis" for its users.[34] Here, the primary subject for improvement is not the human analyst, listening to the ocean. Rather, it is the networked data platform, endlessly modulating to facilitate access and grow new linkages.

If newly present in the formal institutional spaces of the North Coast, Smart Oceans was also broadcasting its vision at local sporting events and coffee shops. Perhaps this is not surprising: as Sevilla-Buitrago observes, the normalizing of enclosure often draws from the so-called free spaces of social reproduction.[35] Media theorists bring a related insight to this idea, namely, that successful integration of networks into new social spaces often depends on their ability to appeal to external activities that networks subsequently seek to govern.[36] According to the energetic ONC reps who staffed a booth at the 2016 All Native Basketball Tournament, ONC's community presentations were being met with considerable local enthusiasm. In Prince Rupert, Dodge Cove, Hartley Bay, and beyond, looming development was a persistent source of anxiety. With its technical sophistication, accommodating staff, and ideological claims to data neutrality, Smart Oceans was resonating in communities eager to find authorities, guardrails, and assurances.

Although they are distinctive institutional trajectories, Community Observatories and Ocean Sense represent interlocking appeals to one of ONC's most purposeful engagement priorities: Indigenous peoples. In its support of North Coast development, the Canadian government was consciously seeking to expand the circulation of fossil fuels and other commodities across unceded Indigenous lands and waters. Efforts to promote capitalist growth here would need to navigate turbulent coastal ecologies and long-simmering community grievances, both of which recommended "mediated forms of . . . recognition and accommodation."[37] An essential part of ONC's appeal to Indigenous peoples is the idea that its network model can incorporate territorial and cultural distinctions. Access to Information documents reveal that ONC had developed a preliminary plan for its "Aboriginal Strategy" in February 2014. The plan specifically touts Smart Oceans for its capacity to feature Traditional Ecological Knowledge—a policy ideal Canada's government agencies have historically failed to realize.[38] ONC reaffirmed this commitment in its "Strategic Plan 2030," which emphasizes "meaningful, ongoing engagement with Indigenous communities," as well as "deep respect for Indigenous knowledge and the communities' rights and title to marine territories."[39] "Over the next decade," the Strategic Plan concludes, "ONC will continue to work with communities to develop ocean monitoring programs, education and training, and youth opportunities that address these priorities."[40]

The informational brochure about the Prince Rupert—Ts'msyen Territory Community Observatory I was given at the 2016 All Native Basketball Tournament had made explicit mention of cultural relationships around culturally valued whales, fish, and territorial waters. Since 2015, ONC has employed First Nations scientists in roles that explicitly seek to nurture inclusionary institutional approaches. It would be a mistake to conclude that these were nothing more than cynical ploys. Multiple individuals I spoke with in Hartley Bay suggested that the council had been impressed by ONC.[41] The presenters who had journeyed to the remote village seemed willing to listen to local concerns in ways government predecessors had evidently not.

In 2017, ONC commenced a subsurface mapping expedition in partnership with DFO and the Council of the Haida Nation, a nation that has championed the pairing of innovative governance technologies with longstanding territorial stewardship commitments. The collaborating team explored SG̲aan K̲inghlas (Supernatural Being Looking Outwards), a seamount and Marine Protected Area located 180 -kilometers off the coast (Haida Traditional Territory), paving the way for ONC's installation of near-bottom moorings and hydrophones on SGaan Kinghlas in 2018.[42] These sensors supported assessment work on variations on ambient sound levels by near and distant shipping events. The collaboration sought to create novel connections between undersea expedition and Haida community knowledge—live-streaming audiovisuals from the seamount for viewers in the Haida communities of Skidegate and Old Masset.

It seems necessary to hold on to two truths here. On the one hand, that projects like the SGaan Kinghlas expedition constitute useful governance initiatives that support local needs. On the other, that they do not reject but perhaps even extend prevailing arcs of development cum modernization. The Chickasaw scholar Jodi Byrd worries about the impact of digital systems where "discrete signs are no longer taken for wonders but are instead taken to signify the system at the level of the micro."[43] What gets lost when contextually formed sounds are aggregated in distant servers in the form of individualized data bits? When territories constituted through clan and even family-specific stories are reprojected in high-resolution spectra? These sensing acts are more than one-off affairs. In a smart ocean, the expanding network is continuously being reproduced as such. It designates social engagements premised on further accumulations of data and sensing. Smart Oceans recognizes that shipping will increase levels of ocean noise around Haida Gwaii. This too proposes the ongoing sensing of spaces like SGaan Kinghlas. Capacities to manage environmental change are predicated not only on capacities to manage data, however, but also on the "algorithmic uncertainties" that surround data.[44] How, I asked Chris Picard at one point, can communities ensure that all this data analysis is not being privileged over actual territorial analyses? It was a question I often pondered during my stopovers in the Hartley Bay band office. As sanctioned by a digitalized state, smartness was generating new kinds of storage and access requirements here. Co-management tasks were literally being downloaded onto overburdened servers and the cell phones of overworked individuals.

Citizen Sensors?

While I was finishing this book, a colleague shared an anecdote with me from his time in the forests of northern Bengal. He reported an exchange involving a camera sensor and three Indian women familiar with the area of the forest where the sensor was installed. Suddenly aware of its presence, one of the women rushes to keep another from speaking. Because the human voice, like the growling tiger, is made up of identifying biometric traits, she had concluded that any kind of communication, whether for community bonding or as a way of alerting predators to their presence, was unsafe.[45] What interests me about the exchange is not the fact that a voice would not have been captured by a camera sensor. Rather, it is the way a form of emplaced sociality was determined to be no longer viable. Like camera traps in northern India, underwater sensors along the North Coast portend the loss of communications that do not have the imprimatur of digital assurance. They portend new confusions between evidence offered by algorithms and cultures rooted in semantic interpretation. On the North Coast, efforts to achieve "marine operational situational awareness" will require not only ongoing accumulations of sensings but sensors as well. Some of these will be fully machinic and positioned in remote locations, but others, like in the India example, will consist in messy overlaps with embodied human forms and practices.[46] As the number of "citizen sensors" contributing to Smart Oceans increases, so too will the amount of data, increasing (it is hoped) its capacity to predict, track, and manage.[47] This does not empower the "citizen sensor" or her capacity to sense. More hydrophones does not translate into augmented listening capacity, but into a series of "partial perceptions" reified into sound clips, usually of thirty seconds or less, which can be aggregated to simulate a collective sensorium.[48]

If the promissory rendering of a predictable North Coast was being engendered by "citizen sensors" and a social acceptance of Smart Oceans, what exactly was being replaced? One object I began to consider more as my time on the North Coast ended was the CB radio—affectionately known in Hartley Bay as the Mickey Mouse. These technologies have long provided a communicative substratum for regional mobilities up and down the region. In 2017, there was still a familiarity to the CB radio. People knew whom to call if a unit needed repairs; they knew from the voice of their interlocutor whether a message had been transmitted or not. Against this, none of the community representatives I talked to would claim any real map of the digital mediations increasingly directing their functional connectivity. Unsurprisingly, I came across other regional media, marked by similar abandonment rhythms: Prince Rupert's small-town movie theater, with its shrinking audiences; Fairview Terminal's public telephone booth, uninstalled in 2015. With the absorptive capacities of the digital network come the unmooring of older ways of knowing and perceiving space, the dereliction of cultural objects that had been socialized through decades of use.[49]

Oceans 2.0: The Data Portal

As noted earlier in this chapter, the sensor data and imagery collected by Smart Oceans' sensors is freely available and searchable through Oceans 2.0 (now Oceans 3.0). This is its platform, an online system that acquires, archives, and distributes observatory data.[50] The data portal is the summa of the Smart Oceans vision: an institutional space where sound, enclosure, and new techno-cultures of digitalization meet. It is also a space in which a new kind of networked sociality resonates, its power and efficiency manifest in digital sounds. By logging on to the Oceans 2.0 landing page in spring 2019 and creating an account, I became part of its network, along with hundreds of other users. I sat there in my Vancouver apartment, alone, using a graphical user interface to scan for hydrophone data. I settled on a node titled Kitamaat Village Underwater Network, in Haisla Territory, at the mouth of the Douglas Channel. Scrolling onto an Ocean Sonics IC-Listen hydrophone hyperlink, I found a promised wealth of metadata: maintenance periods, battery capacities, internal sensor humidity.[51] By selecting a range of recorded dates across a short period, I could listen to brief fragments of digital sound: underwater currents, whale calls, passing vessels. In terms of overall quality, the sounds were not unlike the baseline recordings I had codeveloped in Gitga'at Territory. But as platform designers will aver, the main advantage of these archives is not the sensory experiences they offer, but the statistical analyses they can perform.

Significant for achieving this outcome is the capacity to represent digital sound in spectrogram form. As Joeri Bruyninckx notes, spectral sound data assumes perceptual limits in human hearing. It recognizes that certain phenomenal details captured and recorded as such can appear to humans only in the visual register—that is, in the form of "spacings indicating frequency and coloring suggestive of relative amounts of sound energy."[52] Enabled by "signal transduction" (i.e., the converting of acoustic measurements into visual renderings), I could generate a series of spectrograms through Oceans 2.0, as well as a particular feature called the "spectral probability density map." Spectral probability density maps contain information on ambient noise levels across select regions. This makes them valuable tools for analysis. Passing ships and underwater earthquakes (seaquakes) appear as patterned color sequences: red hyperbolic curves, sharply ascending yellow zags. "Ambient noise measurements are made to assess the acoustic energy underwater from all acoustic sources," an ONC brief explains. "Ambient noise analysis is one of many methods to assess the health of an ecosystem and the anthropogenic impact on it." Moreover, plotting "daily cadences can reveal patterns associated with human activity such as ferries or other regular vessel passages," confirms another brief.[53] Since large vessels typically emit intense underwater noise at low frequencies, they will appear in the shaded region of the spectral plot as a spectral probability density. Darker shadings thus become informational, revealing an acoustic ecosystem in which more "time effort" is exerted at that power level.

Spectral probability density maps—visual outputs realizable from aggregations of sonic data—suggest a logical endpoint of the converging logics of enclosure, digitalization, and smart governance explored in this chapter. Unlike analog sounds, these maps work to realize sonic processes (e.g., the Douglas Channel) in terms of discrete events. For viewers, they carry a long-nurtured scientific fantasy to make sound "tangible and textual by making the invisible visible."[54] But their contemporary use value derives less from measurable human engagement than from the predictive powers they offer governance. The estimation of probability density functions is now a fundamental feature of the computational probabilities that inform much smart environmental governance. Looking at the red and yellow shapes I had generated—from an ocean space near Kitimat—I could only verify what had, in fact, already been determined algorithmically—that is, by pattern-seeking processes enclosed in proprietary algorithms, washing over impossibly large sets of data, giving frequency encodings to information that promised to facilitate subsequent comparisons, analyses, and retrievals.

Digital sound is only one of the streams provisioned by Smart Oceans. It exists alongside captures of underwater temperature, salinity, and dissolved oxygen, all produced to be aggregated and bundled in socially useful ways. But digital sound is a revealing kind of artifact for coming to terms with a smart ocean. In this book's introduction, I offered *sonic capital* as a way of theorizing sound in terms of its inputs to capitalist valorization. I now want to suggest that we can bring sonic capital to the spectral probability density map, as well as to the affective modulations, interpersonal adjustments, and deterritorializations that transpire as communities become enclosed within its smart governance logics. Digital hydrophones generate fragmented data points that can be reassembled into assets. Sensors make predictions that can augment shipping schedules and mitigate shipping risks. Marine governance becomes increasingly coordinated by distant IT firms and corporate interests. To all of this, we can consider the infrastructural lattice of Smart Oceans itself; a growing expression of sonic fixed capital, an infrastructure capable of provisioning new forms of sonic capital. As generated by Smart

Oceans, the spectral probability density map presents sound made useful for economic need; not just shipping (e.g., ocean noise), but all underwater events (earthquakes, tsunamis, submarine landslides, waves, and gas hydrate stability) that might cause disruption, and which require further demands for data amenable to algorithmically enhanced prediction.[55]

As I write these words in 2023, many North Coast shipping lanes have yet to be fulfilled, though several have. But a proof of concept for predictable coastal governance is already near at hand. The example I am referring to manifests south of the region, in the Salish Sea, in the Enhancing Cetacean Habitat Observation (ECHO) project. This is, by some estimates, the largest known nonmilitary vessel noise database in the world.[56] ONC codeveloped this hydrophone-based monitoring system with Port Metro Vancouver and the province of British Columbia, using it to trial dynamic management models that would allow them to optimize marine shipping routes while safeguarding acoustical habitats for whales.[57] Across each lunar month, ECHO generates "ambient noise reports" for its users. These reports allow them to issue modified shipping routes (an adjustment called "lateral displacement") in the face of observed environmental risks. One day, they may provide regulators with data for guiding marine life away from highly trafficked areas in real time, too.[58] If the goal of a Smart Ocean is a predictable North Coast, what we find here is an operational affirmation: an instance of the system learning from distributed networks carrying circulating patterns of human and nonhuman activities, a system seeking to optimize changing regional conditions for the value they can generate.

Conclusion

My last years on the North Coast coincided with the emergence of a new environmental governance ambition, Smart Oceans. Though it is multifaceted in approach and design, its ultimate function, I have suggested here, is to provide a guardrail for "Sustainable Marine Development."[59] For communities looking to safeguard their shores, there are understandable reasons to seek affiliation with Smart Oceans. There are also, as one resident in Dodge Cove candidly put it, "not a lot of alternatives out there." But a smart ocean carries risks. The iteration that seeks to make "Canada's west coast safer for shipping navigation, incident response . . . and tsunami detection" cannot distinguish between the distinctly political natures of the processes (climate change, oil spills, biodiversity loss) it considers.[60] It cannot solve the loss of trust that network thinking (and associated acting) can facilitate. What it can do is enclose spaces and processes in its promise of an endlessly expanding archive, multiplying the points of connection so that events can be better circulated and shared. If anything, it is the anticipation of an event that is the privileged sense of a Smart Ocean, an event made sensible through the state of crisis smartness secretes into everyday ways of seeing, listening, and spatial relating.

But what if the system has no power to sense? "The lack of capacity of the cellular communication system available limited the amount of instruments that could be functioning at any one time," an ONC incident report from 2017 begins, "resulting in either the hydrophone or the video camera transmit [*sic*] data in real time, but not both."[61] Under these circumstances, a smart ocean becomes a space of "original accidents," Paul Virilio's term for the novel disasters technological artifacts can instigate.[62] More than an abstract speculative concern, Virilio's concept has direct relevance to the present discussion. On October 13, 2016, approximately 110,000 liters of diesel fuel spilled into the ocean when the *Nathan E. Stewart* hit rocks ten or so nautical miles west of Bella Bella. Audio recordings obtained by the *Globe and Mail* confirmed that the tiny First Nations community nearby (a village in Heiltsuk Traditional Territory) were the spill's first responders.[63] They arrived on small boats long before the coast guard, but without means to contain the pollutants rapidly spreading across their territory.[64] This sinking recalls another regional historical event. When the four hundred–plus passenger MV *Queen of the North* hit the shoals of Gil Island on a frigid evening in March 2006, it was villagers from Hartley Bay who sped through the watery darkness to rescue people from the sinking ship. As with Heiltsuk, the Gitga'at villagers' efforts suggest the ongoing importance of local knowledge, along with a community's unique capacity to manage its local territory.

Episodes like the *Nathan E. Stewart* and the *Queen of the North* help to explain my emphasis on sociality in this chapter's study of enclosure's normalization. Networked socialities built of "weak ties" seem far less durable when put into action than their animating discourses would suggest.[65] A distrust of weak ties may also help to explain why Janie Wray decided not to connect with Smart Oceans for the rapidly expanding BC hydrophone network she has begun to develop with local communities. It might also explain why certain NGOs, and some First Nations themselves, have agreed to do so only in a limited fashion. This is not to condemn those communities who are collaborating with Smart Oceans. Rather, it is to insist that the ability to retain different ways of doing governance is a critical asset for community decision-making, a point powerfully made in Andrés Luque-Ayala and Simon Marvin's survey of the smart city.[66]

This chapter has found that the enclosure being enacted through Smart Oceans is facilitating the loss of alternative governance cultures on the North Coast. Through digital sound, it has sought to illustrate how this process has proceeded and how it has relied upon a panoply of mediating agents (sensors, territories, human users, etc.). The promise of Smart Oceans is ultimately a utopian one, and utopias, as Fredric Jameson suggests, are valuable ultimately for what they reveal about the limits of future imaginings, as opposed to actual future likelihoods.[67] I suspect the force of my own critique has been limited by many things, not least the contingency of my status as a visitor on the North Coast. By the time I left in 2018, ONC's project had achieved hegemonic levels of consent. It was hard not to connect this to another widely consented-to fact: that LNG, in one form or another, would soon become a reality. I hope I am wrong in my unease. Either way, I suspect the questions that Smart Oceans raised across my time on the coast will be hard to ignore. If anything, we should expect them to circulate through the "resistances and capacitances" of governance, more and more.[68]

CONCLUSION A COUNTRY THAT BELONGS IN NO COUNTRY

Tucked into the middle of the May 1975 issue of the *Native Voice* ("the official mouthpiece of the Native Brotherhood of British Columbia") is a short article about Johnny Clifton, then hereditary chief of the Gitga'at Nation. It concerns Clifton's presence at the United Nations Convention on the Law of the Sea (UNCLOS) meetings in Geneva, Switzerland, and the moment when Clifton rose from the table to thank his fellow attendees—political leaders from Canada and other G7 countries—for the invitation, as "his participation marked yet another step on the road to equality for Canadian Indians." Clifton wanted to "say goodbye through an ancient tribal custom of singing his own song, one composed when he inherited the chieftanship [*sic*] in his settlement near Prince Rupert." With Canada's External Affairs Minister Allan MacEachen beating a tabletop drumbeat at his request, he began chanting—"Hey, Hey, Hey." A solemn Indigenous display all but guaranteed, Clifton went unexpectedly into country music—"Heeey Good Looking/ Whaaa-tcha got cooking"—garnering stunned bewilderment and laughter from across the room.[1]

Clifton's participation at UNCLOS came at a time when oil and gas interests were seeking access to the North Coast—then, as today, through the waterways of Gitga'at Territory.[2] He was not present when the Exclusive Economic Zone frameworks were eventually hammered out at subsequent UNCLOS meetings. For some historians, these represented perhaps the "greatest single enclosure in human history," with nation-states granted sovereign rights over the exploration, extraction, and governance of all living resources within two hundred nautical miles of their shorelines.[3] Against this grim fact, what are we to make of Clifton's act of musical punning? Was

2 — NATIVE VOICE — May, 1975

The Native Voice

The Voice of the Native Canadian
Published monthly by: The Native Voice Publishing Society,
517 Ford Building, 193 East Hasting Street.
Vancouver 4, B.C. Canada
EDITORIAL BOARD: Alvin Dixon, Clarence Joe, Diana Recalma
The Native Brotherhood of British Columbia
ADVERTISING: Rates on request
SUBSCRIPTIONS: $5.00 annually (12 issues)

A Disturbing Position

In a press release (Vancouver Sun, May 6, 1975) George Watts was quoted as saying "Indians to boycott IFAP" which he terms a cover-up for the failure of commercial fishermen's licence limitation scheme. He said IFAP made a few Indian fishermen rich but took jobs from a lot of others. "It put the emphasis on big expensive boats which take most of the catch, leaving less for the small boat fishermen."

The Native Brotherhood of B.C., a self-supporting Native organization, would like to urge B.C. Indian fishermen not to boycott IFAP since this program has proved to be one of the few positive forces in maintaining the Indian fishermen in a competitive position in the fishing industry.

Indians have faced a decline in the number of Native-owned and Native-operated fishing vessels. Three of the main forces for the decline are the commercial fishermen's licensing scheme; restrictions in fishing company rentals, and inflation. The result of these forces is a higher quality of fishing vessel.

Indians are forced to compete at these higher standards on fishing vessels, and IFAP, which was implemented in 1969, has played a positive role in achieving this objective. For instance, IFAP-assisted Indian vessels, which include seiners, gillnetters and trollers, have consistently produced a greater average on gross returns per vessel value than both the other Native-owned vessels, and non-Native owned vessels.

The higher production by the Native vessels provides a greater income for Indians since 95 per cent of their

Was chief's chant 'hey, hey' or 'ha, ha'?

Behind tightly-guarded doors, the cream of Canada's diplomatic community were gloriously duped in Geneva early in May by an Indian chief from Hartley Bay.

Diplomats, ministers in the Newfoundland and federal cabinets, and top government experts all solemnly made believe they were pounding on Indian drums.

This was the rhythmic accompaniment he needed, said Johnny Clifton, president of the Native Brotherhood of B.C., to sing his goodbye to the three dozen other Canadian delegates after eight weeks at the U.N. Law of the Sea conference here.

In a trembling voice, Clifton said his participation marked yet another step on the road to equality for Canadian Indians.

"In 1968, I wasn't allowed to have a drink with you white people, but now I've never seen so much wine and liquor," one of Clifton's victims reported.

His hand quivering from pretended emotion, Clifton said he wanted to say good-bye through an ancient tribal custom of singing his own song, one composed when he inherited the chieftainship, in his settlement near Prince Rupert.

JOHNNY CLIFTON

But he needed the background beat of tom-toms, he said, looking at External Affairs Minister Allan MacEachen and Newfoundland minister John Crosbie.

* * *

While Clifton shouted out the beat, the ministers, the diplomats, the naval commander, the esoteric experts — everyone in the conference room at the delegation headquarters — began thumping hands on their knees or the table.

It was a highly-charged emotional moment as Clifton accelerated the beat, grunting, "Hey, hey, hey, hey. . ."

And then he broke into the traditional tribal song.

"Hey, hey, good-looking, whatcha got cookin'. . .?"

It was the best laugh the diplomats had had since a treasury board bureaucrat in Ottawa had okayed mid-conference flights home — so long as they squeezed a return transatlantic trip and the home visit into one weekend.

* * *

Clifton, hereditary chief of the Hartley Bay band 80 miles south of Prince Rupert, said on his return: "Everything was going so seriously I thought I might just change the atmosphere before I left."

He said he hoped no one was offended by the practical joke and didn't know how reporters learned of the incident, "but I really suckered them into it."

Clifton, a fisherman, was among B.C. Native Brotherhood delegates seeking to protect Canadian fishing interests at the conference. He said he could not comment on the proceedings.

— Peter Calami
Southam News Service
Vancouver Province

FIGURE C.1. Article by Peter Calami, "Was Chief's Chant 'Hey, Hey' or 'Ha, Ha'?" *Native Voice*, May 1975, File GR-1738.110.25—Native Voice, Royal BC Museum, Victoria, BC.

a message being communicated about the processes underway? Was it remorseful, ironic, an incipient instance of Indigenous refusal politics? Or was it just a joke from someone who likes Hank Williams? And then there is the fact of Hank Williams; not just any country musician, but one whose incomparable songwriting and Indigenous heritage has endeared him to so many Native communities.[4]

Today, country music remains a remarkably fecund site from which to probe the politics of identity in North America, and the North Coast as well.[5] Country music, with its "dual dialectic of memory/forgetting and sentimentality/amusement," establishes a field of significations that has historically appealed to a range of emplaced listenerships.[6] In 2008, Geoff Mann argued that country music calls white Americans to their whiteness, its characteristic twang serving as a reminder of the less multicultural, more authentic homeland so many white Americans evidently crave.[7] In response, Kristina Jacobsen found that the same country songs can call Indigenous people "with equal force to working-class Indigenous identities."[8] Nadine Hubbs, for her part, seeks to restore consideration to the Mexican contributions buried in the history of this musical form, as part of a broader project of grasping country's significance to queer culture.

Of all the music I heard on the North Coast—in cars and aboard boats; along the Dodge Cove docks and in Hartley Bay kitchens—it was country that I heard most often. I didn't realize until much later how, while never fully registering as a subject of interest, country music was "music in the background"—to quote Adorno.[9] Differentiated by generation and style and subgenre (Merle Haggard vs. Alan Jackson vs. Florida Georgia Line, say), it was nevertheless the common thread among the Ts'msyen elders convening at Tim Hortons for the millionth time that week; the aging fishermen across Chatham Sound; the young men driving Alberta pickup trucks—all of a piece with it, all listening to versions of country music.[10] Country music would carry the seeds of newer, possibly more critical, listening communities too. In spring 2018, I helped organize a country music tour featuring Saltwater Hank (aka Jeremy Pahl), talented local musician Apples (Jessica Rampling), and Simone Schmidt, an artist of international renown. For six days we toured Haida Gwaii. I stood in the back of the old village halls and gymnasiums, grinning as the musicians took turns harmonizing and sharing works. Jeremy introduced a song about salmonberry picking (titled "Liimi Mak̲'ooxs") that showcased his emerging work on Sm'algyax language revitalization. Simone outlined a powerful new album about incarcerated women in a nineteenth-century Ontario asylum (*Audible Songs from Rockwood*). Apples sang about the Sangan River on her adopted home of Haida Gwaii. For these musicians, "country belongs in no country."[11] It is not about the assurances of a white middle-class existence. It does not need to pay fidelity to the genre's problematic tropes. And yet somehow, each of these artists will manage to channel country music's most singular capacity: to figure belonging—for all the mysteries and complexities belonging can entail.[12] In this sense, perhaps the music of Saltwater Hank is of a piece with the song of Johnny Clifton in that Geneva boardroom. In responding to MacEachen, maybe Clifton was, in fact, just doing what he said he would do: reconnect with his home.

To begin this conclusion with country music affords me a chance to say something about another kind of "country" project. What about Canada? Another version of *A Resonant Ecology* might have been a more sustained intervention into the question of national identity and statecraft, with North Coast development as a story of a broader "technological nationalism," or perhaps as a moment in Indigenous efforts to "unsettle Canada."[13] My decision to eschew a methodological nationalism was guided by a few factors: the globally scaled issues of capital and development, debates in sound studies and eco-Marxism, and the North Coast as unceded Indigenous space. Above all, I wanted to showcase the singularity of the region in question. This is, in many ways, a region that belongs to no country too. It is a social ecology composed of different kinds of rhythms, histories, and agencies. This is affirmed by the humpback whales who journey from Taylor Bight to Oahu every fall. It is also affirmed by the other residents who hold the movements of the coast to be the source of their lifeways, cultures, and renewals.

Singing Wire and Hydrophone

In 1999, John Durham Peters suggested that "the succession from the 'singing wire' through the microphone, telephone, and phonograph to radio and allied technologies of sound marks perhaps the most radical of all sensory reorganizations in modernity."[14] In its engagements with hydrophones, noise maps, and black metal MySpace pages, this book elaborates on Peters's insight. It considers how new sonic technologies across the North Coast point to new spheres of politics, species entanglements, and sensorial change. Through sonic capital, *A Resonant Ecology* links proliferating sound technologies to new frontiers of capital, including a smart governance that now encompasses various geographies with assetizing surveillance. But truly innovative acoustical science is also taking shape. Combining music theory, AI, and movement ecology as the basis of new kinds of conservation science, exciting new experiments are positing sound as a critical input for determining the thresholds at which animals respond to sudden anthropological and ecological changes.[15] These endeavors are relevant for reimagining sonic politics and the architectures of environmental governance they may one day inform. For geographers and sound theorists alike, following the coproductions of sound, science, and governance could expose fruitful lines of multidisciplinary research.

Another of this book's prevailing interests is with development. From hydrophone hisses to distorted guitars, *A Resonant Ecology* seeks to affirm how the microtemporalities of sound persistently relate broader development forces, including the *longue durée* structures of settler colonialism and capitalism. More than twelve years after the Fairview Terminal expansions were announced, Prince Rupert's dilapidated streetscape looks much the same. And yet in significant ways, the city and the broader region have witnessed profound reorganizations across this span—of governance regimes, political dynamics, marine ecologies, and cultural forms. In fall 2024, LNG Canada is set to commence sending tankers down the Douglas Channel, positioning it to become the largest greenhouse gas emitter in all of British Columbia. For this project, at least, the "coming boom" has finally come, even as energy markets and climate pledges give it little chance of becoming the panacea the Government of Canada had promised. There are many things to be said about development, but of particular interest for me here has been its sense-making dimensions. If development always entails multiple perceptional adjustments, development will seek to organize this assortment around its guiding telos, its status as an unfolding reality. As a common sense, development can be a tremendously powerful orienting force. Once consummated as collective narrative, the "coming boom" can be a hard thing to "un-sense."

To interrogate development, *A Resonant Ecology* emphasizes sound's status as an uncertain object of social mediation. The mediation concept is not a magic key that can address all the ontological and political horizons raised by the case studies. But it does offer a useful means for grasping interconnection and change, and for storytelling different encounters in terms of the socio-technical conditions, discursive logics, and material practices that shape them. Hopefully, the arguments explored in this book can support interrogations that attend to new kinds of sonic mediation—including those that remain "unfamiliar or uneasy on the ear."[16]

In chapter 1, I considered encounters with whale song at Cetacea Lab and how changing ways of listening at the station heralded broader transformations to institutional models of regional conservation. In chapter 2, I reversed the lens to consider killer whales as listeners and outlined the process by which ocean noise risk became an assistive pretext for new markets in data sensing. Chapter 3 considered the interplay of state space and settler grammars of place, and how one coastal community struggled to negotiate the noisy politics of port restructuring. In chapter 4, I considered the significance of Indigenous black metal to the politics of Indigeneity on the North Coast. I argued that Gyibaaw's music sheds light on the negotiated aesthetic forms that mobilize different kinds of musical listenership while affording powerful personal connections to place. Finally, chapter 5 explored the ascendance of Smart Oceans, a governance project that draws vast spaces and life-forms into new circuits of measurement and enclosure, with implications on the way particular data forms, like digital sound, engender and inform sociality. Smart Oceans points to an emerging global governance ambition: sound's implication within new fields of ecological forecasting and movement ecology, alongside capacities for predictive tracking and social surveillance.[17]

Over the last two decades, patterns of degradation—from rising levels of ocean acidification, to incidents of red tides and algae blooms, to accelerating rates of glacier loss—have become more common in the North Coast. The cases featured in this book speak to the multitudinous ways ecological crisis became sensible to communities on the North Coast as an ongoing series of calamities and incidents, rather than an always-deferred future happening. These are ecological effects of development, to be sure, but in their socialization, they are also artifacts speaking to the troubling ways we measure. To a distressing degree, we remain within the impasse Geoffrey Bowker identified in 2009, when he observed that "while computer memory is happily outpacing Moore's law, the information carrying capacity of the earth (in the form of genetic information in species) is diminishing at an inverse exceptional rate."[18] Bioacousticians have for several years now spoken of "acoustic fossils" of functionally extinct biota, preserved for posterity in digital form.[19] How to move from data to action? If research on the North Coast shares in a general concern, the best kind also looks to imagine tangible solutions. A good example is a 2023 paper led by Eric Keen, which finds that the LNG projects poised to feature in Gitga'at Territory could result in as many as eighteen humpback whale collisions per year ("ship strikes"), before outlining a set of measures that could reduce this figure.[20] "We're dreading it," Janie Wray remarks in a *Guardian* news article about the study, "because we know we're going to be the ones that are going to witness it."[21] In one sense, Janie returns us to where we started; to anticipatory listenings and the coming boom (if with new spider webs of networks and sensors). Perhaps new "sonic refugia" will emerge in the coming years, as was the case with humpback whales in the early 2000s.[22] But there is little doubt the forecast for the coast is grim, perhaps characteristically so.

Is there any hope for truly progressive environmental governance on the North Coast? If so, what will ensure it can resist the multiphase enclosures of Big Tech and the capitalist state and support truly alternative models of science and multispecies well-being? With her Whale Sound project, Janie Wray seems dedicated to uncovering an answer. Along with research partners in Hartley Bay, Klemtu, and Bella Bella, Janie is seeking to enact different social relations of cetacean acoustic monitoring.[23] "We're really trying to listen collectively," she explained to me in 2021, "to do the work slowly, and to [both] protect and share what we find." Supporting this effort is BC Whales, Janie's new research lab on Fin Island (Gitga'at Territory). BC Whales embraces new and experimental research tools. It actively seeks scientific knowledge production and collaboration of the sort Cetacea Lab sometimes strained to resist. Both the Fin Island station and the Whale Sound network are uncertain ventures, fraught with institutional and ethical challenges. But both recognize the need for new kinds of governance—experimental, ethical, and locally attuned—against the repetitions of technocracy and knowledge enclosure.

A Resonant Ecology began as an effort to heed Jonathan Sterne's call to engage the "sonic imagination," to forge a new kind of development critique through the interests, insights, and questions of sound studies.[24] Along with this, it proposed a critical geographical sensibility, one attuned to the unevenness and variegation of the development process. If stories of sonic culture remain stories of geographical encounters, then it is important "to historicize the specific forms that the making of natures takes, and to be able to do this in geographically situated ways."[25] As I embarked on my research, it became evident that still other orientations would have to guide it. The Indigenous place-thought I encountered in Hartley Bay, while not an active site of exploration in this book, provided me with crucial grounding. Place-thought would become important to my sense of this project's analytic limits. Part of the value of a partial engagement, as Marxist-feminist thinkers have long observed, is the opportunity to reflect on the partiality of knowledge, and so provide openings for new insights.[26] Perhaps regional work more attentive to cosmology, the aural encodings of gender, and the interplays of sound and ability can draw from what has been offered here. Certainly, the North Coast offers many materials for attentive listening.

As was the case 150 years ago, the formative political logics here remain capitalism and colonialism, whose combination continues to enable and delimit possibilities for listening and sound-making. In considering the North Coast as an archive of responses to capitalist-colonial development, at several points of this book I returned to the legacies of salvage anthropology. The possessive logics that mark sound's manifestation in digital form can be traced back to desires first articulated during the salvage period, when anthropologists like Franz Boas saw it as necessary to extract Indigenous sounds from their webs of relations. Recent years have seen a flurry of Indigenous responses to this legacy, spanning questions of museum curation to experimental poetry. For Robin Gray, "rematriation," grounded in Indigenous law and led by Indigenous women, names a growing regional effort to recover "stolen, misappropriated, or commodified belongings"—including sounds.[27] From this perspective, institutions must go beyond repatriation to achieve true reparation and redress. They must reconsider provenance over ownership, access, and control, along with the fact that songs and oral traditions are, for Ts'msyen like Gray, always grounded in Ayaawx (laws). Rematriation is a powerful way of figuring sound studies as a kind of resistant practice. As an audiopolitics, it is multigenerational in scope—reaching back to the voices of elders and ancestors while seeking to entrust future generations with vital cultural materials. As such, it infuses many of the moments considered in this book—like the recording of whale song or the recording of acoustic baselines—with new questions, and new possibilities for alliances. What would it mean for settlers to de-possess the songs that they have taken from these lands and waters—from Indigenous musicians, but also from whales and rivers and wolves? To what horizons might this lead? How might rematriation shift "common sense" at a time of relentless extraction and enclosure? If Michael Denning is correct in his assessment that "the decolonization of the territory was made possible by the decolonization of the ear," perhaps an Indigenous-led rematriation could lead to truly empowering ways of collective listening, rooted in collective desires to value the world differently.[28]

Thank You for an Unexpected Visit

There is no end to the story of a resonant ecology, only a succession of soundings, echoes, and refrains. There are more places to discover, and as a *k̠'amksiwah* only on the North Coast a brief moment, so much I will never know. But since all books must end, let me close this with one more country song from a truly singular place. It was already dusk when our boat arrived at the seaweed camp tucked deep inside the archipelago. It was early spring, and children were still playing by the shore, but the lights twinkling from inside the shacks gave the suggestion that most of the activity had already moved inside. I walked up the beach and followed my hosts through a door, becoming engulfed in a haze of cigarette smoke and fish smells. Several people were sitting around a table. An elderly woman stirred a bubbling pot with a huge hunting knife. There was a song on a radio at the back of the room, past the rows of tackle and dried halibut. It was the unmistakable voice of Roy Orbison:

> Oh, sweet dream baby
> Yeah sweet dream baby
> Sweet dream baby
> How long must I dream?

It was all too intoxicating: the smells and the smoke; this extraordinary community in this extraordinary place. And that voice. What was it doing here? In my perplexity, I was too late to observe the setup, the quieting of the chatter, the roomful of gazes now upon me. One of the seated figures, a great fisherman who had long intimidated me, eyed me sternly before breaking into a grin: "*You!* What are you doing here?"

NOTES

INTRODUCTION

1 The key initiative supporting these efforts was the Asia-Pacific Gateway and Corridor Initiative. See Government of Canada, "Canada's Asia-Pacific Gateway and Corridor Initiative." For nationally and regionally focused coverage, see Stalk and McMillan, "Seizing the Continent"; Ircha, "Serving Tomorrow's Mega-size Containerships"; Markey and Heisler, "Getting a Fair Share." The historical nature of this ambition is captured in Hick, *Hays' Orphan*; and Large, *Prince Rupert*.

2 Dembicki, "At Ground Zero for Next Huge Enviro War."

3 The Great Bear Rainforest agreements were formulated during the same period the Asia-Pacific Gateway and Corridor was taking shape. For critical accounts of the rainforest agreements, see Dempsey, "The *Politics of Nature* in British Columbia's Great Bear Rainforest"; and Low and Shaw, "Indigenous Rights and Environmental Governance." For a commentary that considers the intercession of oil and gas, see Rossiter and Burke Wood, "Neoliberalism as Shape-Shifter." A general, albeit self-promoting, institutional history of the agreements can be found at Coast Funds, "From Conflict to Collaboration."

4 I follow local convention in specifying the Ts'msyen Nation as the general political organization to which broad ideas of culture are assigned, and Gitga'at First Nation as one of the territory-specific bands constituting the Ts'msyen Nation. Other bands include Gits'ilaasü, Gitsumkalum, Kitasoo/Xai'xais, Metlakatla, Lax Kw'alaams, and Gitxaala.

5 There is too much literature to recount here, but one unofficial (i.e., white settler) account that Spencer himself recommends is Miller, *Tsimshian Culture*.

6 Throughout, the capitalist-cum-colonial development in question here is the sort Gillian Hart terms "Little d"—that is, development "as geographically uneven but spatially interconnected processes of creation and destruction." See Hart, "D/developments after the Meltdown," 119. References to sustainable

marine development and economic development throughout should be read with Hart's idea in mind.

7 See Ritts and Bakker, "Conservation Acoustics."

8 Morton, *Dark Ecology*, 99.

9 A longer history of this story would engage the forces that have historically accorded sound a secondary status under modernity's hierarchy of the senses. See Sterne, *The Audible Past*, especially the classic passage on the "Audio/Visual Litany" in the book's introduction.

10 Marx, *Economic and Philosophic Manuscripts*, 114.

11 Here, I am thinking of a tradition spanning Simmel, "The Metropolis and Mental Life"; Adorno, "On the Fetish-Character of Music and the Regression of Listening"; Bull, "No Dead Air!"; Drott, *Streaming Music, Streaming Capital.*

12 Ghosh, *The Great Derangement*, 5.

13 See especially Morton, *Dark Ecology*. For a good example of Morton's approach in action, see the multidisciplinary Art-Science collaboration by Sonic Acts and Hilde Methi, *Sonic Acts–Dark Ecology*, which involved various in situ listening exercises across damaged Eurasian ecologies between 2014 and 2016. Also significant is the work of Bruno Latour. In his influential *Facing Gaia* lectures, Latour brings a distinctly techno-managerialist perspective to this idea, writing: "This is what it means to live in the Anthropocene: 'sensitivity' is a term that is applied to all the actors capable of spreading their sensors a little farther and making others feel that the consequences of their actions are going to call back on them, come to haunt them." See Latour, *Facing Gaia*, 141. Both tendencies—Mortonian and Latourian—persist across popular works of sound studies. For examples, see LaBelle, *Sonic Agency*; Bakker, *The Sounds of Life*.

14 The specific phrase can be found in Friedner, *Sensory Futures*, but my uptake of the idea draws more from Bratton, *The Stack*.

15 Popova, "Nature Is Always Listening."

16 TallBear, "An Indigenous Reflection on Working," 232.

17 Malm, *The Progress of This Storm*, 173. Nature and society are coproduced forms. The separation of nature and society, as proposed here, thus proceeds on an analytical basis. This move allows for the consideration of how each historically distinct stage of society consists in its distinctive metabolic exchanges with nature. See Malm, *The Progress of This Storm*, 61; and for another helpful elucidation, see Saito, *Marx in the Anthropocene*, 110.

18 The term *sonic materialism* has been invoked in various ways in sound studies, most notably by Christoph Cox, who uses it to develop a Deleuzian approach to sound study. See Cox, *Sonic Flux*. Another lineage extends from Salomé Voegelin, who uses *sonic materialism* to construct dialogues with the work of Quentin Meillassoux (e.g., Voegelin, "Sonic Materialism"). As this introduction hopefully makes clear, my uptake is more firmly located in debates that extend from the central claims of historical materialism.

19 LaBelle, *Acoustic Territories*, 298.

20 There are too many texts to list here. Some of the pieces that particularly influenced the writing of this work are Gallagher, "Field Recording"; Jasper, "Sonic Refugia"; Gallagher, Kanngieser, and Prior, "Listening Geographies"; MacFarlane, "Governing the Noisy Sphere"; Kanngieser, "Sonic Colonialities"; Revill, "How Is Space Made in Sound?"; Hemsworth, "'Feeling the Range'"; Dawkins and Loftus, "The Senses as Direct Theoreticians"; Lally, "Policing Sounds."
21 The term *sonic agency* comes from LaBelle, *Sonic Agency*. My usage is slightly different from LaBelle's. I reference agency as involving capacities to act, in this case in relation to sound, rather than proposing sonic agency as a novel theoretical construct.
22 The "Litany" is first proposed in Sterne, *The Audible Past*, 15.
23 Bakker, *The Sounds of Life*, 14.
24 Greco, "Hear That?"
25 James, *The Sonic Episteme*.
26 Williams, *Culture and Materialism*, 8.
27 Adorno, *Essays on Music*.
28 Guthman, *Wilted*. Guthman's book is exemplary in its combination of assemblage and classic political economy approaches to the study of nature. It is capacious without stretching the analytic too far—as is the case, for instance, in Richard Grusin's "radical mediation," which conceives mediation as a "fundamental process of human and nonhuman existence." See Grusin, "Radical Mediation," 125.
29 Harvey, "On the History and Present Condition of Geography," 6.
30 See Heinrich, *An Introduction to the Three Volumes of Karl Marx's Capital*.
31 Robinson, *Hungry Listening*.
32 Boas's 1898 essay "On Alternating Sounds" might be the inaugurating effort of twentieth-century sound studies. In this short piece, Boas discusses his efforts listening to Ts'msyen speakers, noting how their word for *fear* produced a variability of written transcriptions—*päc* and *bas*, in this case. For Boas, this indexed a more general linguistic inaccessibility regarding Ts'msyen soundings. The upshot was not to abandon ethnographic study but to account for cultural variability in the collection effort. For more on Boas's text, see Fee, "Rewriting Anthropology and Identifications on the North Pacific Coast."
33 McCarthy, "Limits/Natural Limits," 419.
34 Harvey, "The Enigma of Capital and the Crisis This Time," 90.
35 Steingo and Sykes, *Remapping Sound Studies*, 12.
36 Simpson, *Mohawk Interruptus*, 113.
37 Mann, "From Countersovereignty to Counterpossession?"
38 Various uses of the term *sonic capital* are in circulation today. See Bürkner and Lange, "Sonic Capital and Independent Urban Music Production"; Kerr, "Compression and Oppression"; Schulze, "Das sonische Kapital"; Hracs, Seman, and Virani, *The Production and Consumption of Music in the Digital Age*. All share a general interest in linking sound—and usually, digital sound—to contemporary expansions of the capitalist value form. I use the term in a

rather limited fashion, specifically, as a way to chart new inputs into the capitalist valorization process.

39 The idea of linking sonic materialism and sonic capital emerged out of conversations with Sumanth Gopinath. The framework is indebted to the extensive knowledge Sumanth has shared over the years regarding interplays of sound and logics of capital (as well as via his landmark text *The Ringtone Dialectic*). My account of an absent-present sonic capital also draws from Michael Heinrich's *An Introduction to the Three Volumes of Karl Marx's Capital*, with its interest in depicting the fetishistic idea of capital as a force with productive powers all its own, independent of labor, and expressive of an "overwhelming social interaction that cannot be controlled by individuals." See Heinrich, *An Introduction to the Three Volumes*, 12.

40 To my knowledge, there has yet to be a full-fledged study of the ways that aboveground industrial development conditions are supporting today's environmental noise sectors. Various kinds of noise assessment, mitigation, and forecasting routinely transpire around industrial megaprojects, sounding out the limits of the economy (to paraphrase Stefan Helmreich), determining the spaces, niches, and moments where sonic capital can circulate amid expansive and unruly spaces of change.

41 Attali, *Noise*.

42 The policy was most explicitly laid out in the 2016 Oceans Protection Plan. See Government of Canada, "The Prime Minister of Canada Announces the National Oceans Protection Plan."

43 Halpern and Mitchell, *The Smartness Mandate*, 15.

44 See, for example, Cruikshank, *Do Glaciers Listen?*; Gordillo, *Rubble*; and Ogden, *Loss and Wonder at the World's End*.

45 Following local convention, I use the term *North Coast* in this book, cognizant of the many other names that have been used to describe the general region in question.

46 Ricketts, *Ed Ricketts from Cannery Row to Sitka, Alaska*.

47 See, for example, Kahn, *Noise, Water, Meat*.

48 Zylinska, "Hydromedia," 45.

49 Following convention, as well as the self-identifying language of the Indigenous peoples I encountered, I use the broad signifier *Ts'msyen* to describe the cultural group that encompasses several of these nations: Metlakatla, Gitga'at, Gitxaala, Kitasoo/Xai'xais, Gits'ilaasü, Gitsumkalum. Gitga'at thus appears as a specific instance of a broader Ts'msyen culture.

50 Braun, *The Intemperate Rainforest*, 231. "Purification machine" is Braun's term for those places of "un-civilization" where settlers commit acts of self-reinvention through nature.

51 Barman, "The West beyond the West."

52 See, for example, McAllister and McAllister, *The Great Bear Rainforest*.

53 On the Guardian Watchmen Program, see Coastal First Nations, "Eyes and Ears of the Land and Sea."

54 For a related discussion of this idea focused on other contexts, see Kanngieser, "Sonic Colonialities."

55 Arguably, the inaugurating sound of settler colonialism and its accompanying capitalist political economy came a century earlier, in the form of the gun salutes issued by James Colnett's English scouting ships in the 1780s. See Galois, *A Voyage to the North West Side of America*.

56 Barbeau, *Pathfinders in the North Pacific*, 209–10. Duncan's relationship to the duties expected of him by the Church of England was complex, but as a missionary he consistently required that his Ts'msyen congregations relinquish their Indigenous traditions in the process of their inevitable assimilation into a larger North American society. See Neylan, *The Heavens Are Changing*. Neylan's book is a wonderful account of the brass band culture through which Christian conversion efforts were routinely negotiated by Indigenous peoples on the North Coast. It serves to establish an additional point: Duncan was by no means the only local missionary who combined his acoustic proclivities with his settler colonialism. When he wasn't helping to coercively secure the purchase of Ts'msyen reserve land to the Grand Trunk Railroad, Archbishop Frederick Du Vernet (1860–1924) was perfecting Radio Mind, a telepathic transmission system that enabled listening to distant speakers by harnessing vibrations moving through air. For a history of this lesser-known but also momentous religious figure, see Klassen, *The Story of Radio Mind*.

57 The territorialization of bells in relation to Catholicism has done similar work for centuries, as Alain Corbin has observed. See Corbin, *Village Bells*.

58 One of the most compelling recent investigations into the political economy of settler colonialism in Canada is Pasternak, *Grounded Authority*.

59 From an economic and ecological standpoint, the most important sector in the development in question is liquefied natural gas. By the early 2010s, Kitimat and Prince Rupert had become the ports of choice for a number of projects hoping to leverage market conditions and the region's geographical proximity to Asia. See Petroleum Human Resources Council of Canada, *Labor Demand Outlook for BC's Natural Gas Industry*.

60 Large, *Prince Rupert*. Consider the city's founding poem ("Prince Rupert") by Riddell Elliot, published in the local newspaper, the *Evening Empire* (March 28, 2014): "Behind you, young Prince Rupert / Rests a country big and grand / The raw material for an Empire / Wooing the magic touch of man. / Here in you progressive Rupert, / Is the gateway big and bold / To a kingdom of vast riches—/ Of copper, silver and pure gold."

61 The combined North Coast and Lower Mainland regions probably contain more acoustic telemetry than anywhere else in the world. See Hussey et al., "Aquatic Animal Telemetry"; Favali, Beranzoli, and De Santis, *Seafloor Observatories*.

62 Bridge and Perreault, "Environmental Governance." The authors argue for a definition in which environmental governance is concerned with the ways different entities—such as actors, spaces, processes—can be "brought into durable forms of alignment" (485).

63 The foundational accounts here are Coulthard, *Red Skin, White Masks*; Simpson, *As We Have Always Done*.

64 Simpson, *As We Have Always Done*, 173.

65 Duffus and Dearden, "Non-consumptive Wildlife-Oriented Recreation."

66 Large, *Prince Rupert*.

67 Ochoa Gautier, *Aurality*, 3.

68 The most egregious example probably came from the province of British Columbia under Christy Clark (2011–17). Her administration's promise of a $100 billion Prosperity Fund that would ensue from the LNG buildout was never remotely realistic.

69 Critique of the project from coastal communities drew from Enbridge's arrogant dismissals of local territorial concerns. This was brought home in the Douglas Channel Watch's (2014) tragic-yet-funny "disappearing islands" campaign. See Bowles and MacPhail, "The Town That Said 'No.'"

70 Perhaps the most notable land defense—at least in terms of the capacity to draw international attention—is the ongoing Unis'to'ten Camp (Wetsu'wet'en Territory, 2008–).

71 In July 2012, I first contacted the Gitga'at Nation to express my interest in supporting their resistance to the Enbridge Northern Gateway project. I proposed interviews with local elders whose deep knowledge of the land could help elucidate the threats the shipping prospects posed. The proposal was rejected. If I wanted to work with the Gitga'at, research director Chris Picard explained to me, there needed to be "tangible benefit" to the community, over and above good messaging. Between 2013 and 2016, I worked with the nation to position and collect data from eight SM2 sensors in select areas along the shipping route that cuts though the middle of the nation's marine territory. The eco-acoustics baseline project aimed to support the nation's territorial protection efforts and provide scientific materials for intervening in a state-directed energy project assessment process (the Joint Review Panel). See Ritts et al., "Collaborative Research Praxis to Establish Baseline Ecoacoustics Conditions in Gitga'at Territory."

72 De Leeuw, Cameron, and Greenwood, "Participatory and Community-Based Research," 188.

73 Benjamin, *Illuminations*.

74 Adorno, *Essays on Music*. See also Leppert, "Music 'Pushed to the Edge of Existence.'"

75 Born, "On Musical Mediation."

76 Mills, "Deaf Jam," 25.

77 Moore, *Capitalism in the Web of Life*.

78 Lefebvre, "Space and the State" in *State, Space, World*; Goeman, *Mark My Words*.

79 Martineau and Ritskes, "Fugitive Indigeneity."

80 Hall, "Gramsci's Relevance"; Coulthard, *Red Skin, White Masks*.

81 Sevilla-Buitrago, "Capitalist Formations of Enclosure"; Andrejevic, "Ubiquitous Computing and the Digital Enclosure Movement."

1 Dobell, "Caamaño."
2 See Pilkington, Meuter, and Wray, "Written Evidence Submission to the Joint Review Panel."
3 Observed numbers of humpbacks peaked in 2015. Since that time, the year-to-year totals have varied.
4 Critical Whale Habitat is an official Government of Canada designation ostensibly granting species protections. See Department of Fisheries and Oceans, "Species Listing under Canada's Species at Risk Act."
5 Fletcher, "Gaming Conservation," 155.
6 Adorno, "Late Style in Beethoven." My reading has also been influenced by the Richard Leppert–edited *Essays on Music*; Spitzer, *Music as Philosophy*; Spencer, "Lateness and Modernity in Theodor Adorno"; and Wheeldon, *Debussy's Late Style*. I have taken liberties in reading this literature into whale music, proposing a listener-centered model of late style.
7 Adorno had the late classical music of Beethoven in mind when he proposed late style. My uptake vis-à-vis whale music is less invested in moments of "intransigence, difficulty, and unresolved contradiction" (Said, *On Late Style*, xiii), though I do emphasize the formally complex elements of whale song over and against the popular comparisons with Muzak. We should also note that the period of whale music's emergence witnessed a range of successful nature music albums that incorporated similar experiments with media technologies. These include Irv Teibel's *Environments* (1969), the first in a long series devoted to looped nature sounds. Others include Newmark and Gimbel, *Ambience One* (1970), billed as "an Adventure in Environmental Sound," and Dan Gibson's *Solitudes* (1981–94), which pioneered new techniques of field recording. See, for example, Gibson, *Solitudes—Volume 1*. Clearly, a commodification of nature proposed choices within a burgeoning music consumerism, with different subgenres newly available to individuals looking to express their social identities. For an excellent history of New Age, see Szabo, *Turn On, Tune In, Drift Off*. On the Adorno connection, see also Leppert, "Introduction."
8 Said, "Thoughts on Late Style." Said helpfully highlights the temporalizing dimension in his discussion of Adorno's concept.
9 Mowitt, *Percussion*, 217.
10 Connor, "Menagerie of the Senses," 11.
11 Connor, "Menagerie of the Senses," 12.
12 I have developed a strange fascination with these tales. Among the titles one would have to count as solidly in this genre are Weyler, *Song of the Whale*; Payne, *Among Whales*; Morton, *Listening to Whales*; Visser, "Killer Whales in New Zealand Waters"; Kelly, *Song for a Whale*; Siegel, *Whalesong*; Picoult, *Songs of the Humpback Whale*; Chapin and Forster, *Sing a Whale Song*; Sheldon, *The Whales' Song*; Wilson, *The Longest Whale Song*; Tardif, *Whale Song*; and Johnston, *Whale Song*.
13 McVay, "The Last of the Great Whales," 2.

14 A. Trites, personal communication with author, March 13, 2013.

15 At the same time, Hermann and Janie embarked on parallel research projects to understand killer whales'—and later, fin whales'—spatial-acoustic needs. The Northern Resident Killer Whales Cetacea Lab also monitored in depth calls at much higher frequencies (typically, 4–18 kHz) than those of humpbacks. They are much shorter in duration as well.

16 According to cetologists, humpback whales produce a "complex, structured, series of vocalizations called song." See Ford et al., *Passive Acoustic Monitoring of Large Whales*, 7. Humpback whales are one of only three species on the planet known to alter their vocal displays actively and repeatedly (the others are humans and bowhead whales).

17 Whitehead and Rendell, *The Cultural Lives of Whales and Dolphins*, 43. Only male humpbacks are known to sing. Until recently, they typically did this in the winter breeding season. Cetacea Lab would observe some of its most intense song displays in the summer and fall.

18 For an overview, see Whitehead and Rendell, *The Cultural Lives of Whales and Dolphins.*

19 In *The Sounding of the Whale*, D. Graham Burnett estimates that over 150 items of popular and classical music—from Alan Hovhaness to Kate Bush—have sampled or thematized whale phonation since 1970 (634n). Classic uptakes of humpback song include George Crumb's *Vox Balaenae*; Paul Horn's *Haida and Paul Horn*; and Paul Winter's *Common Ground.* There is also Pink Floyd's *Meddle* (1972), the centerpiece of which is a trippy humpback-inspired guitar solo in the middle of the twenty-two-minute epic "Echoes." We should also recognize that killer whale iterations of whale music effectively moved from a narrowly environmentalist pursuit into something tourists and publics could routinely encounter at institutions like Sealand of the Pacific, the Seattle Aquarium, the Vancouver Aquarium, aboard a growing number of hydrophone-toting whale watching boats, as well as in the local record store.

20 Birdsong also exists in the music idiom but does not have the same cultural modifiers attached as regards nonintelligence. In this case, it is the sonorous quality, less the evidence of humanlike culture, that motivates the musical association.

21 Payne and McVay, "Songs of Humpback Whales," but as evidence of the incredible longitudinal interests of Katy Payne, see also Payne and Payne, "Large Scale Changes over 19 Years in Songs of Humpback Whales."

22 Katy Payne, interview with author, April 24, 2020.

23 Bateson, "The Logical Categories of Learning and Communication."

24 Meintjes, *Sound of Africa!*

25 Rothenberg, *Thousand-Mile Song*, 1. The significance of the BC coast in this regard is well known. This is the setting where Greenpeace was founded in the early 1970s alongside a suddenly flourishing Save the Whales movement, where the worldwide whale oceanarium trade was started, the anticaptivity movement after that, and a whale tourism industry that now grosses in the

hundreds of millions. See Hoyt, *Orca*; Zelko, *Make It a Green Peace!*; Weyler, *Greenpeace*; Francis and Hewlett, *Operation Orca*; and Colby, *Orca*. Whale song's symbolic power as musical environmentalism acquired an ersatz emotionalism in the New Age uptakes of song in the late 1970s and 1980s. The liner notes to a New Age whale music album by Terry Oldfield (written in 1993) are representative: "There have been many whale-fusion albums since *Out of the Depths* . . . but few duplicate its ethereal majesty. And while the sense of ecological loss is always palpable, Oldfield turns it into a meditation on the frailty and temporality of all life, not just that of our oceanic brethren" (Oldfield, *Out of the Depths*).

26 Cvetkovich, *Depression*, 4.

27 McVay, "The Last of the Great Whales"; Mustill, *How to Speak Whale*, 9. In his 1995 retrospective *Among Whales*, Roger Payne finds present-day whale song to be lacking the power and clarity it once had. Notably, Payne humanizes whale song ("the quality of many whale notes is the same as the quality of human notes"), considers a universal musical Platonism, and professes human industrialization as a cause of its qualitative decline. For Payne, the sounds we hear today are imbued with loss—a sonorous reminder of a once-vibrant ocean. See Payne, *Among Whales*, 147.

28 I found little reason to doubt the sincerity of intent here—from the introductory refrain told to all interns ("The lab only operates here because of the continued generosity of the Gitga'at Nation") to Hermann and Janie's various efforts to share data with the Band Council. Throughout my time, radio exchanges between Cetacea Lab and passing Gitga'at marine vessels—the Gitga'at Guardian Watchmen, in particular—would occasionally pierce the headphoned listening bubbles, confirming our status as uninvited guests working to support a broader conservation project.

29 For a more measured account, see Morton, *Listening to Whales*. See also various episodes in Howard White's edited classic the Raincoast Chronicles, in particular Norris, *Time and Tide*. There are also useful anecdotes in Francis and Hewlett, *Operation Orca*; and Colby, *Orca*.

30 Born, "On Musical Mediation." Born suggests that "Adorno's aversion to interrogating the specific institutional arrangements of . . . music" limits his analysis (15).

31 Spong's work converting Greenpeace to the whale cause is explored at length (if also with a degree of glorifying excess) in Weyler, *Song of the Whale*.

32 A sentence written by Spong summarizes the desired transition nicely: "I dropped my posture of remoteness, opened my mind, and personally engaged myself in Skana's learning." Spong, "Introduction," 7.

33 Spong, "Whale Communication," 184.

34 Smith, *Uneven Development*, 27–29.

35 Leppert, *The Sight of Sound*, 29.

36 Jenemann, *Adorno in America*.

37 The unlikely beginnings of PAM are in naval bioacoustics. See Ritts and Shiga, "Military Cetology." The insights cetologists can draw from PAM continue

to widen. As one history of the practice notes, "the amount of information transferred by sound is for cetaceans completely unknown and is subject to scientific debate." See Zimmer, *Passive Acoustic Monitoring of Cetaceans*, 41.

38 The reference here is to the classic account of the "audio-visual litany" as featured in Sterne, *The Audible Past*, 7–9.

39 Years later, scientists would complain of the "self-noise" of the icListens. This has motived new coast-wide investments in JASCOS AMAR ($50,000.00) hydrophones, the Cadillac of hydrophones, according to one senior scientist. I discuss this in greater detail in chapter 2.

40 With their expanded frequency ranges and sensitivities, digital hydrophones can facilitate new understandings of the locations and distributions of song, as well as observational data on a number of observed animal behaviors (such as calls expressive of animal foraging versus traveling). Another added benefit is spatial: whereas single hydrophones can record sound arriving from any direction, arrays of hydrophones can be deployed to construct an observational network. Incoming signals can then be manipulated to listen in any direction with even greater sensitivity than a single hydrophone element, as well as to triangulate signals to better track particular sound sources.

41 Another issue was that the analog hydrophones had a limited dynamic range, meaning Cetacea Lab was "missing out on some of what is going on in certain frequencies where whales vocalize" (Janie).

42 With their data interoperability, scaling possibilities, and storage demands, digital hydrophones also result in reams of new data-based work, eventually producing new jobs beyond the station for storing and managing its terabytes of sound.

43 Mercado and Perazio, "Similarities in Composition and Transformations of Songs," 42.

44 I personally encountered more than a few instances where a whale call corresponded to only the faintest visual impression on a spectrogram. If not for the experience and presence of Hermann and Janie, who could draw from their considerable acquired experience—the way the upswing sound of an R Clan killer whale call "peeked out" from background noise differently than a G Clan call, for instance—I would have had no way of giving a confident determination that the call had taken place.

45 Barua, "Animating Capital."

46 Althusser, *On the Reproduction of Capitalism*.

47 See MTV, "Pipeline Wars."

48 Igoe, "The Spectacle of Nature in the Global Economy of Appearances."

49 Debord, *Society of the Spectacle*.

50 Chion, *Music in Cinema*.

51 For good overviews on the logics of spectacle shaping the interplay of environmental conservation and neoliberalism (and which lay a critical basis for the present account) see, in particular, Igoe, "The Spectacle of Nature in the Global Economy of Appearances"; Igoe, "Nature on the Move II"; Büscher,

"Nature 2.0"; Sullivan, "Beyond the Money Shot"; and Fletcher, "Taking the Chocolate Laxative."

52 See Wray and Meuter, "Project Cetacea Lab."

53 Tsing, *Friction*, 11.

54 Fletcher, "Gaming Conservation," 155.

55 Tsing, *Friction*, 4.

56 I am not at liberty to requote these exchanges, but a close analog is at hand. OrcaLive, which has operated out of nearby OrcaLab (Hanson Island, BC) since 2001, enables distant listeners to dial in to OrcaLab's sounds and partake in its listening activities—much like Cetacea Lab's Listen to Whales function. In her ethnography of OrcaLive, "Watching the Whale Watchers," Donna Bray notes several respondent statements whose discursive coordinates remain illustrative:

> D.Listener: "Wow, what a show we are having. Thanks Orca Live. What a way to work. Typing, watching, and listening to orcas. Made my day." spokes@Anacortes, Wash, 29 July.
>
> HappyHost: "It's amazing!! If someone told me a year ago I would be able to recognize certain pods I would have laughed. I know A's for sure. But I can't distinguish which group of [Killer Whale clan] A's yet. Now I know I's. Are those resting calls? I just love hearing them."

57 Kareiva, "Ominous Trends in Nature Recreation," 2757.

58 See Affolderbach, Clapp, and Hayter, "Environmental Bargaining and Boundary Organizations."

59 See Büscher and Fletcher, *The Conservation Revolution*. Some of this discourse would find its way into so-called new or Anthropocene conservation (Nordhaus and Shellenberger, "The Long Death of Environmentalism").

60 Personal field notes taken from across several visits in summer 2015 substantiate this observation.

61 Bakhtin, *The Dialogic Imagination*.

62 "The steeper walls of bare bedrock and narrow channels of the interior fjords, secluded as they are from major shipping lanes, may offer an attractive acoustic space," Keen, "Whales of the Rainforest," 107. Although Keen's statement pertains to fin whales, it applies well to humpback whales, whose low-frequency calls might also represent efforts to source the North Coast's reflective and (still) considerable propagation conditions. In conversation, Keen suggested as much to me, though he was careful not to suggest there was confirming evidence. His geological insight may also carry some applicability to the higher-frequency-calling killer whales. A well-cited paper by Whitehead and Ford, "Consequences of Culturally-Driven Ecological Specialization," for instance, considers how the culturally transmitted ecological specializations noted of killer whales have key spatial determinants.

63 By one estimate, the total is 9,400 large whales; see Nichol et al., *British Columbia Commercial Whaling Catch Data*.

64 Gumbs, *Undrowned*, 11.

65 The breakup was explored in a 2018 documentary filmed at Cetacea Lab. See Leuze, *The Whale and the Raven.*

66 Spencer, "Lateness and Modernity in Theodor Adorno."

2. VALUE IN INJURY

1 George, *Ninety Percent of Everything*, 3.

2 UNCTAD, *Trade and Development*, 4–5.

3 Hesse and Rodrigue, "The Transport Geography of Logistics and Freight Distribution."

4 McDonald, Hildebrand, and Wiggins, "Increases in Deep Ocean Ambient Noise."

5 Lütticken, *History in Motion*, 126.

6 Helmreich, *Sounding the Limits of Life.*

7 Erbe, MacGillivray, and Williams, "Mapping Cumulative Noise from Shipping."

8 Duarte et al., "The Soundscape of the Anthropocene Ocean."

9 Besky and Blanchette, *How Nature Works*, 19.

10 Sunder Rajan, *Lively Capital.*

11 Gieryn, "Boundary-Work." This also helps to explain its wide epistemological inheritance: an ocean noise that gathers insights from oceanography, physical acoustics, cetology, marine planning, even transportation engineering.

12 See, for example, Gedamke et al., *Ocean Noise Strategy Roadmap*; Duarte et al., "The Soundscape of the Anthropocene Ocean."

13 Lave, Mirowski, and Randalls, "Introduction," 668.

14 Mills, "Deaf Jam."

15 The phrase is from Pugliese, *Biopolitics of the More-Than-Human.*

16 Johnson, "At the Limits of Species Being," 283, emphasis added.

17 Mills, "Deaf Jam," x.

18 Quoted in Mulvin, "Talking It Out," 8.

19 As suggested by this reference, the model of mediation given in this chapter is one of assemblage, that is, assemblage as a "complex and dynamic process whereupon the collective's properties exceed their constitutive elements." See Guthman, *Wilted*, 12.

20 A good example here is AP Moller-Maersk's so-called Radical Retrofit. In 2017, the world's largest shipping conglomerate, AP Moller-Maersk, retrofitted five of its large container ships, finding that reducing propeller cavitation decreased low-frequency sound pressure levels by 6–8 dB while improving fuel efficiency. See Gassman et al., "Underwater Noise Comparison."

21 See, for example, Alger, Lister, and Dauvergne, "Corporate Governance and the Environmental Politics of Shipping."

22 Ritts and Bakker, "Conservation Acoustics."

23 Loon, *Risk and Technological Culture*, 2.

24 Heise and Barrett-Lennard, "The Calm before the Storm."

25 Farmer, "Acoustical Studies of the Upper Ocean Boundary Layer," 315. For more on this subject, see Urick, *Principles of Underwater Sound.*
26 Since the 1950s, an acoustic-informatic approach to marine space has remained dominant for both the oceanographic community and many others. For work in this area, see especially Shiga, "Sonar and the Channelization of the Ocean"; Camprubi and Hui, "Testing the Underwater Ear"; and Han, "Precipitates of the Deep Sea."
27 For a paper that speaks to the continued significance of these early efforts, see Farmer, Vagle, and Booth, "A Free-Flooding Acoustical Resonator."
28 I am not suggesting that acoustical understandings of the ocean began in the 1960s. There are longer histories. Rather, my point is that the discourse of ocean noise was largely a World War II/Cold War phenomenon.
29 Oreskes, *Science on a Mission*, 387.
30 See Dow, Emling, and Knudsen, "Survey of Underwater Sounds, Sounds from Surface Ships."
31 Wenz, "Ambient Noise in the Ocean"; and Chapman and Price, "Low Frequency Deep Ocean Ambient Noise."
32 National Research Council, *Present and Future Civil Uses of Underwater Sound.* Cetology's sonic interests began decades earlier, but the 1970s were the inflection point.
33 Collard, *Animal Traffic*, 21.
34 Miles and Malme, *The Acoustic Environment and Noise Exposure.*
35 National Research Council, *Ocean Noise and Marine Mammals*; World Conservation Congress, "Undersea Noise Pollution"; and International Whaling Commission, "Report of the Standing Working Group."
36 Landecker, "Postindustrial Metabolism," 496.
37 See Lehman, "A Sea of Potential."
38 Galison, "Trading Zone."
39 Risk was already common in proceedings leading up to the Vancouver conference, such as this one from the National Marine Fisheries Service: "It was agreed by all that the primary reason for the assessment of risk, particularly as related to marine mammals . . . is communication between those with expertise in business, scientific, and engineering disciplines and those relying on such expertise to make decisions." Carlson, de Jong, and Dekeling, "Workshop One," 658.
40 McWhinnie et al., "The Grand Challenges in Researching."
41 For examples, see Battistoni, "Bringing in the Work of Nature"; Barua, "Animating Capital"; and Collard, *Animal Traffic.*
42 Barua, "Animating Capital," 655.
43 National Research Council, *Marine Mammal Populations and Ocean Noise*, xi.
44 Horowitz and Jasny, "Precautionary Management of Noise," 226.
45 Weilgart, "The Impacts of Anthropogenic Ocean Noise."
46 For representative work over this period, see Holt, Noren, and Emmons, "Effects of Noise Levels and Call Types"; Lusseau et al., "Vessel Traffic Disrupts

the Foraging Behavior"; Williams et al., "Behavioural Responses of Male Killer Whales"; Williams et al., "Severity of Killer Whale Behavioral Responses"; and Erbe et al., "Identifying Modeled Ship Noise Hotspots."

47 See Dempsey, "Biodiversity Loss as Material Risk." Protections under Canada's Species at Risk Act identify killer whales as what Dempsey calls a "regulatory risk"—that is, a risk producing new costs to development actors owing to the possible biological impacts.

48 Baumgartner et al., "Persistent Near Real-Time Passive Acoustic Monitoring," 1477. Exemplary here is the Sound and Marine Life JIP, a multimillion-dollar, multiyear research program funded by large oil and gas companies in collaboration with academics at universities and research institutions. JIP has provided over US$55 million in research funds and supported the publication of dozens of peer-reviewed manuscripts on the subject of underwater sound. See Isaksen, "Advances to the Science of Sound and Marine Life," 57.

49 See the Canada's Ocean Supercluster website, https://oceansupercluster.ca.

50 Lawson and Lesage, "A Draft Framework to Quantify and Cumulate Risks."

51 See NRDC, "Sounding the Depths."

52 These certifications are available from a wide range of organizations, including the Agreement on the Conservation of Cetaceans of the Black Sea, Mediterranean Sea and Contiguous Atlantic Area and the Joint Nature Conservation Committee. For a discussion, see David, *Progress Report on the Implementation of an ACCOBAMS Certification*.

53 Harris et al., "The Challenges of Analyzing Behavioral Response Study Data."

54 David Johnstone, personal communication with author, March 4, 2013.

55 See Clark et al., "Acoustic Masking in Marine Ecosystems."

56 University of St. Andrews, "New Tool to Assess Noise Impact."

57 A good history of this project can be found in Lehman, "A Sea of Potential."

58 Baumgartner et al., "Persistent Near Real-Time Passive Acoustic Monitoring."

59 Agardy et al., "Mind the Gap."

60 Ambach, "Current and Proposed Projects with Underwater Noise Implications."

61 Added to this is another challenge: the largest vessels in the region at the time were Panamax class and Capesize class. Increased regional growth could see much larger vessels, possibly in the VLCC class (upward of 250,000 deadweight tonnage). Because noise output is closely correlated to vessel size, different ship-size scenarios could project different noise levels.

62 Southall et al., "Marine Mammal Noise Exposure Criteria."

63 Enbridge, "Northern Gateway Application," 10-39, 10-51.

64 NRDC, "Submission of the Natural Resources Defense Council"; Nowlan et al., "WWF-Canada Submission."

65 McCauley, *Review of Documents Associated with Assessing Environmental Impacts*, 4. See also NRDC, "Submission of the Natural Resources Defense Council."

66 Department of Fisheries and Oceans (DFO), "Species Listing under Canada's Species at Risk Act."

67 In 2019, amendments to the Canada Shipping Act gave Transport Canada more ability to regulate ocean noise for small vessels—such as whale watching boats. But the familiar issue of national-scale regulation returns: commercial fleets derive their license and mandates from the IMO and do not feature here.

68 It has become apparent that this effort cannot scale and does nothing to reduce the long shelf life of many currently operating vessels, some of which are planned to run for another forty years.

69 Another review of ten major international agreements has similar conclusions (Duarte et al., "The Soundscape of the Anthropocene Ocean," 8). The exception is the European Union's *Marine Strategy Framework Directive*, which was enshrined in 2008 and applies only to the North Sea.

70 DeCola et al., *Nuka Research and Planning Group*, 75.

71 Erbe and Farmer, "A Software Model to Estimate Zones of Impact."

72 Cusick, "Towards an Acoustemology of Detention."

73 Mills, "Deaf Jam."

74 Besky and Blanchette, *How Nature Works*, 19.

75 For example, Bijsterveld, *Sonic Skills*.

76 Chua, "Containing the Ship of State," 273.

77 Puar, *The Right to Maim*, xv–xvi.

78 Birtchnell, Savitzky, and Urry, *Cargomobilities*.

79 The ship was seized by creditors for unpaid bills. See Lough, "Hanjin Scarlet and Crew Detained."

80 Philippine Overseas Employment Administration, "Deployed Overseas Filipino Workers."

81 Sekula, "Fish Story," 582.

82 Williams et al., "Quiet(er) Marine Protected Areas," 158. As the authors conclude, "Given the expensive and logistical challenges associated with moving shipping lanes, we see value in pointing out opportunities where maintaining the status quo would offer conservation benefit with little socioeconomic cost."

83 Weilgart et al., "Signal-to-Noise." See also Whitehead and Weilgart, "Marine Mammal Science." For work that elaborates on the scientific concerns around noise, see Weilgart, "The Impacts of Anthropogenic Ocean Noise."

84 Here as well, one is advised to consider Sunaura Taylor's important call for an "environmentalism of the injured." See Taylor, *Beasts of Burden*.

85 Sunder Rajan, "Epilogue," in *Lively Capital*, 448.

3. "PORT NOISE"

1 See chapter 2 of this book.

2 See Thompson, *The Soundscape of Modernity*; Radovac, "The 'War on Noise.'"

3 Noise abatement campaigns often exemplify the sorts of moral panics that arise when environmental issues are compounded by class tensions. For

some excellent treatments that also pursue a political economy approach, see Radovac, "The 'War on Noise'"; MacFarlane, "Governing the Noisy Sphere"; Thompson, *The Soundscape of Modernity*; Schwartz, *Making Noise.*

4 Goeman, *Mark My Words.*

5 Lefebvre is a wonderfully productive thinker for exploring relations between space, sound, and economic power. To my knowledge, the geography literature contains no other uptakes of sound studies that utilize his "state space" concept. But other approaches that hold some insight to this chapter's efforts include Lefebvre's "rhythmanalysis" analytic; his contributions to notions of an Urban Sensorium, and the filiation of Lefebvre with critical phenomenologies of space. For examples, see Kinkaid, "Re-encountering Lefebvre"; Goonewardena, "The Urban Sensorium"; and Palmer and Jones, "On Breathing and Geography." See also Kipfer and Goonewardena, "Urban Marxism and the Post-colonial Question."

6 Lefebvre, *State, Space, World*, 238.

7 Lefebvre, *State, Space, World*. See also Lefebvre, *The Production of Space.* Everyone I spoke to in Dodge Cove agreed that the PRPA, a recognized state institution governed under the Ports Act, was the primary engine of local economic development. Of course, this helps to explain its powerful symbolic role as bearer of local economic change.

8 Byrd, *The Transit of Empire.*

9 Daloz, *We Are as Gods*, 7.

10 Goeman, *Mark My Words*, 236.

11 This is discussed in Vimalassery, Pegues, and Goldstein, "Introduction."

12 Vimalassery, Pegues, and Goldstein, "Introduction."

13 Goeman, *Mark My Words*, 11.

14 Neil Brenner, introduction to Lefebvre, *State, Space, World*, 21. As Brenner notes, Lefebvre's most seminal works were tightly connected to concerns with a "Fordist Marxism." This means new dynamics must be considered under neoliberalism. For more on my reading of Lefebvre, see also Charnock, "Challenging New State Spatialities"; and Akhter, "Infrastructure Nation."

15 Unsurprisingly, the full implications of the colonial relationship in North America escaped Lefebvre's critical grasp. His writings were formed within decidedly Eurocentric contexts but remain useful for tracing out the spatial dimensions of shifting relationships to state power in many contexts. Lefebvre left ample room for considering state space alongside notions of territory, central to settler colonial workings. For example, Lefebvre writes, "Is not the secret of the State, hidden because it is so obvious, to be found in space? The State and territory interact in such a way that they can be said to be mutually constitutive" (*State, Space, World*, 228). See Kipfer and Goonewardena, "Urban Marxism and the Post-colonial Question."

16 Here I wish to emphasize that in the present case state space does not get reworked though noise per se, but rather through the various practices, technical forms, and institutional procedures its emanations *entail*. These include

novel opportunities to map space, quantify natural features, administer population flows, and proliferate new social relations of discipline, governance, and expertise.

17 Goeman, *Mark My Words*, 237.

18 When I met Lou, she was in the middle of editing a book anthology bearing this moniker, a vivid essay collection of firsthand recollections from women who came to the coast from Vancouver, California, eastern Canada, and beyond. See Allison and Wilde, *Gumboot Girls*.

19 Blanchet, *The Curve of Time*, 80.

20 Blanchet, *The Curve of Time*, 80.

21 McCartney, "'How Am I to Listen to You?'"

22 Riley, "Digby Disturbed." Riley's article was about opposition to the Aurora LNG plant that had been proposed for the southern tip of the island.

23 SQCRD, "Dodge Cove Official Community Plan," vi.

24 I was disappointed with my overall returns: only 15 percent of Graham Ave. residents responded, and less than 10 percent of the Water St. residents did. But in Dodge Cove, 66 percent of residents responded, which is itself revealing of the political ethos and aesthetic sensibilities of the community.

25 See Radovac, "The 'War on Noise'"; Thompson, *The Soundscape of Modernity*. Another rich example of this critique in geography is MacFarlane, "Governing the Noisy Sphere."

26 "Unwanted sound" being perhaps the most common descriptor for noise.

27 Large, *Prince Rupert*.

28 See Meggs, *Salmon*.

29 Bowman, "Rail Operations Declined after Waterfront Fire," n.p. Two of the oldest Prince Rupert individuals I interviewed, Don Scott and Bill McNish, could recall the noise of common steam locomotives—the N-5-a Consolidations model and the H-10-a Ten Wheelers—carrying loads of halibut and salmon into the Rupert waterfront in the 1950s.

30 Gardiner, "Railroads Built This Country."

31 See Ommer, *Coasts under Stress*; Halseth and Ryser, "Rapid Change in Small Towns"; and Wilson and Summerville, "Transformation, Transportation, or Speculation?"

32 These developments pertained not just to the port expansion but also to the proposed Aurora LNG. Still, community members saw them as interconnected elements in the regional industrial development process and so combined them forcefully in noise.

33 The context is further detailed in an article in the *Northern View*. Campbell, "Friction Grows over Dodge Cove Community Plan."

34 Later, I would learn that this "free pass" had an additional grievance: unlike the struggling residents on either side of Chatham Sound, the wealthy Port of Prince Rupert pays no taxes.

35 The insight resonated with many people I spoke to, including Prince Rupert city official Zeno Krecic, whom I met to discuss implementations of the city's

noise ordinance. ("How are we going to implement?" Zeno responded. "We don't have a pot to piss in!")

36 Keystone Environmental Ltd., *Environmental Screening Document.*

37 Keystone Environmental Ltd., *Environmental Screening Document,* 113.

38 Brian Denton, letter to Health Canada, 2011. Personally shared with author.

39 Environmental Assessment Office, "Collected Public Comments for Aurora LNG."

40 Bosworth, *Pipeline Populism,* 51.

41 See Prince Rupert Port Authority, "Notice of Intent to Pass Whistle Cessation Resolution." Personal copy of Brian Denton.

42 Hacking the Mainframe chat forum, accessed July 21, 2015, https://forum.hackingthemainframe.com.

43 Wilson and Summerville, "Transformation, Transportation, or Speculation?," 228.

44 Brown, *States of Injury,* 52–77.

45 For example, Thoreau, *Walden*; and Schafer, *The Tuning of the World.*

46 Whitehouse, "Listening to Birds in the Anthropocene"; and da Luz, "Undocumented Birds."

47 Minutes, "Aurora LNG—Dodge Cove Community Meeting—Feb 17th, 2014, 7 P.M." These were contained in a blue binder held in the village community hall.

48 SQCRD, "Dodge Cove Official Community Plan," viii.

49 See Robinson, *Hungry Listening.*

50 See Budd, "'The Story of the Country.'"

51 Vimalassery, Pegues, and Goldstein, "Introduction."

52 Blanchet, *The Curve of Time,* 80.

53 Ames, *The North Coast Prehistory Project*; MacDonald and Cybulski, "Introduction"; and MacDonald and Inglis, "An Overview of the North Coast Prehistory Project."

54 Archer, "Village Patterns and the Emergence of Ranked Society"; Coupland, Stewart, and Hall, *Archaeological Investigations in the Prince Rupert Harbour Area.*

55 This is an idea that regional archaeologists, such as David Archer, have also considered. See, for example, Archer, "Village Patterns and the Emergence of Ranked Society."

56 Letham et al., "Assessing the Scale and Pace of Large Shell-Bearing Site Occupation."

57 Byrd, "Follow the Typical Signs."

58 During the time of my visits to Dodge Cove (2012–17), the North Coast saw three Indigenous land defenses (functioning as blockades) erected in opposition to regional development—including one very close to Dodge Cove at Lelu Island (the other two, Unis'to'ten Camp and Maddii Lii, featured in inland territories). In Dodge Cove, people I spoke to recognized these conflicts as involving points of shared concern—but not as constituting "the same fight."

59 Perhaps unsurprisingly, the solidaristic period of alliances referred to here was much more complicated than is often remembered, particularly as it pertained to Indigenous-settler relations. On this, see especially the work of Charles Menzies. Menzies, "Us and Them."

60 Byrd, *The Transit of Empire.*

4. *ANCESTRAL WAR HYMNS*

1 For good critical context on the Stephen Harper quote ("Canada has no history of colonialism"), see Walia, "Really Harper, Canada Has No History of Colonialism?"

2 Fanon, *The Wretched of the Earth*, 47.

3 Fanon, *The Wretched of the Earth*, 64.

4 See Coulthard, *Red Skin, White Masks*, especially chapter 5: "The Plunge into the Chasm of the Past: Fanon, Self-Recognition, Decolonization."

5 Vizenor, *Manifest Manners*, vii.

6 I am grateful to the community scholar and wonderful poet of the North Coast Sarah de Leeuw. It was Sarah who first told me about Gyibaaw.

7 Denning, *Noise Uprising.*

8 Most of my fieldwork with Gyibaaw consisted of the antidisciplinary ethnographic ritual of hanging out. I had an audio recorder and a few preset questions, but things never happened in sequence. Interviews with Jeremy and Spencer often devolved into digressions spanning kung-fu movies, Ts'msyen history, and the powerful shared Indigenous-Jewish love of Canadian Chinese food.

9 Martineau and Ritskes, "Fugitive Indigeneity," 4.

10 Simpson, "Paths toward a Mohawk Nation," 112.

11 Simpson, *As We Have Always Done*, 198.

12 Barbeau and Beynon, *Tsimshian Narratives.*

13 Martineau and Ritskes, "Fugitive Indigeneity." See, for instance, Hagen, "Musical Style, Ideology, and Mythology"; Berger, *Metal, Rock, and Jazz*; Kahn-Harris, *Extreme Metal*; Moynihan and Søderlind, *Lords of Chaos*; Morton, "At the Edge of the Smoking Pool of Death"; Noys, "'Remain True to the Earth!'"; and Lukes, "Black Metal Machine."

14 Hegel, *Aesthetics*, 518.

15 Cole, *Captured Heritage.*

16 Barbeau, "Songs of the Northwest," 104. It is important to note that some anthropologists admitted to struggles with their sense making and even wrote reflexively about the limits of their analytic tools. "In closing," Ida Halpern wrote in 1976, "I would like to suggest that our terminology of 'meaningless, nonsensical' should be changed to 'enigmatic' when dealing with syllables of whose meaning we are not, at the time, cognizant." See Halpern, "On the Interpretation of 'Meaningless-Nonsensical Syllables.'"

17 Rifkin, *Beyond Settler Time.*

18 Hall, "Black Diaspora Artists in Britain," 3. Hall has a nuanced attention to the way "moments" form from contexts defined by different modes of production, historical processes, geographies—all finding expression in distinctive "articulations." This is useful for pursuing the idea that a music may have anticipated—even fomented—some of the cultural forces that have surged in the wake of neoliberal globalization, including Indigenous resurgence. At the same time, the account is insufficient for appreciating how Indigenous musical creativity emerges in relation to specific places, a point emphasized in Martineau and Ritskes, "Fugitive Indigeneity."

19 With his "grounded normativity," Glen Coulthard considers the creative and ethical frameworks mediated by Indigenous place-based practices. See Coulthard, *Red Skin, White Masks*; and Coulthard and Simpson, "Grounded Normativity."

20 Noys, "'Remain True to the Earth!'"; Hagen, "Musical Style, Ideology, and Mythology"; Karjalainen and Kärki, *Modern Heavy Metal*.

21 Rousseau, *The Reveries of the Solitary Walker*, 42. Consider the following from noted black metal singer Varg Vikernes (aka Burzum): "The modern man has lost his connection to the soil of his forefathers. The modern man's connection to his forefathers and the gods of his blood is lost too. He travels all across the Earth as a creature with no roots anywhere" (in Moynihan and Søderlind, *Lords of Chaos*, 23). It is not hard to read the exaggerated statements of Vikernes, who served multiple years in jail in the 1990s for murdering his ex-bandmate, as a tragic expression of Jamesonian "blank parody." On this, see Jameson, *Postmodernism*.

22 Morton, "At the Smoking Edge of the Pool of Death."

23 Quoted in Springer, "A Conversation with Dr. Keith Kahn-Harris."

24 For an illustrative example, see Karjalainen and Kärki, *Modern Heavy Metal*.

25 Noys, "'Remain True to the Earth!,'" 105.

26 Kahn-Harris, *Extreme Metal*, 5.

27 Berger, *Metal, Rock, and Jazz*.

28 Vizenor, *Manifest Manners*, vii.

29 See *Encyclopaedia Metallum*, "Browse Bands—Alphabetically," https://www.metal-archives.com/lists. Bands upload their own biographies and thus have the opportunity to self-identify as Indigenous.

30 Hagen, "Musical Style, Ideology, and Mythology," 187.

31 *Metal Nation*, "Understanding Black Metal."

32 Coulthard, *Red Skins, White Masks*, 109.

33 The fourth member of Gyibaaw, guitarist Brandon Dyck, self-identifies as white. In its effort to explore themes of Indigeneity, this account downplays Brandon's contributions to the band.

34 Innes and Robertson, *Indigenous Men and Masculinities*, 3.

35 Fanon, *The Wretched of the Earth*, 40.

36 Such spaces represent a deadly calculus in the lives of young Indigenous women—as local Indigenous community aid groups have long noted. But problems for Indigenous men have long been rife as well.

37 McCreary, "Mining Aboriginal Success."

38 Before Gyibaaw relocated to Prince George, Jeremy and Spencer used to practice in a trailer in Burns Lake. As Cherill Pahl told me (in what would be one of the most enjoyable interviews I've ever conducted), "I used to hear them just banging away in that trailer. I had no idea it was music. It was just noise, noise, noise. But they were having so much fun."

39 Prescott-Steed, "Frostbite on My Feet," 47; Ishmael et al., *Helvete 1*.

40 TallBear, "An Indigenous Reflection," 232. For a rich sound studies exploration of these themes (from which my account also draws), see Ochoa Gautier, *Aurality*.

41 Tomlinson, *The Singing of the New World*.

42 But only on the cassette version.

43 Whyte, "Indigenous Science (Fiction)," 228.

44 See Roth, *Becoming Tsimshian*; Miller, *Tsimshian Culture*.

45 Lacan, *Ecrits*, 719. See also Dolar, *A Voice and Nothing More*.

46 It should be noted that Indigenous protocol remained a characteristic feature of Gyibaaw's restorying approach, even as their music departed from convention. To research "Nekt," the band contacted members of the Gitksan Nation, who oversee the warriors' story, and asked permission to explore the topic. They incorporated story details from living Gitksan members—including the family of Gyibaaw bassist Norm McLean. As Jeremy and Spencer would tell me, the Nekt story had been symbolically buried by the Canadian Heritage plaque that now adorns the site and makes insufficient mention of the area's true identity.

47 Gyibaaw, "Tour Statement." A poster copy was personally shared with me by Jeremy Pahl.

48 It should be stressed that land acknowledgments in 2010 were not performatively and institutionally absorbed by the state in the way they became by the end of decade. See Robinson, *Hungry Listening*.

49 Rutherford, "Wolfish White Nationalisms?" The settler colonial relation to wolves persists in more material ways too. Many people in Hartley Bay were enraged when, in 2016, British Columbia sponsored a wolf cull that resulted in the deaths of 163 animals in the northeastern part of the province. See Weichel, "163 Wolves Killed."

50 Spencer noted with great pride how biologists have come to recognize the wolves of the North Coast (and Central Coast to the south) as one of the most unique wolf subgroups in the world. For biologists, coastal wolves form a sophisticated society whose isolation from human predation has evidently endowed them with group characteristics unlike those seen anywhere else. They confidently swim through ocean water, scavenge for salmon and shellfish, and communicate in a spectacular fashion across island distances. For an example, see Muñoz-Fuentes et al., "Ecological Factors Drive Differentiation in Wolves from British Columbia."

51 Barbeau and Beynon, *Tsimshian Legends*.

52 Freccero, "Wolf, or Homo Homini Lupus," 92.
53 Hall, "Race, Articulation, and Societies Structured in Dominance."
54 An exposé of the Inquisition encounter in *Decibel Magazine* featured a lengthy interview with Daniel Gallant, who would become a major source of controversy in the black metal community. See Norton, "Ex-Skinhead." For more details of the controversy, see the reply in *Decibel Magazine*: Mudrian, "Inquisition Frontman Dagon"; see also *Shamelessnavelgazing*, "Inquisition and Black Metal's Fascism Problem"; and Scholars from the Underground, "Prince George Indigenous Band." For some public reflection from Jeremy and Spencer, specifically in their iteration as the Saltwater Brothers, see CBC, "Saltwater Brothers."
55 Robinson, *Hungry Listening*, 17. Here Robinson suggests that "the public capacity to hear resentment and repulsion remains questionable."
56 In March 2018, I attended a wonderful, visceral, Vancouver performance by the Black feminist, Indigenous Australian doom metal group Divide and Dissolve. Midway through the set, Glen Coulthard tapped me on the shoulder and said, "This is the antidote to the problem in your chapter." Thank you, Glen.

5. SMARTEST COAST IN THE WORLD?

1 Gillespie, *Custodians of the Internet*, 1.
2 Haiven, *Art after Money, Money after Art*, 160.
3 Digital platforms have become a productive site for geographical inquiry in recent years. As many scholars have argued, digital platforms are more than neutral knowledge forums. Rather, they are key socio-technical intermediaries with the capacity to shape more and more political and social life. See, in particular, Langley and Leyshon, "Platform Capitalism," 11.
4 ONC, "Products and Services Deliverables Report," 000198.
5 Halpern and Mitchell, *The Smartness Mandate*, 14.
6 The quote is from Transport Canada, "Ocean Networks Canada."
7 ONC, "Products and Services Deliverables Report," 000198.
8 For interconnected examples, see Bakker and Ritts, "Smart Earth"; and Ritts and Bakker, "Conservation Acoustics."
9 The literature on marine enclosure is extensive. See Ritts and Simpson, "Smart Oceans Governance," for a summary of some recent work.
10 Halpern and Mitchell, *The Smartness Mandate*, 28.
11 Sevilla-Buitrago, "Capitalist Formations of Enclosure." Numerous geographers have developed enclosure as a spatial logic intrinsic to the movements, value practices, and dispossessive actions of contemporary capitalism. Enclosure is a historical phenomenon and contemporary social process, encompassing both the accelerating acts of contemporary privatization and the walling-in of English common areas in the fifteenth and sixteenth centuries.
12 Andrejevic suggests that "digital enclosure" combines "spatial characteristics of land enclosure with the metaphorical process of information enclosure"

("Surveillance in the Digital Enclosure," 27). See also Andrejevic, "The Pacification of Interactivity"; and Andrejevic, "Ubiquitous Computing and the Digital Enclosure Movement." This approach is useful for relating digital enclosure to the historical enclosures facilitated by Canada's settler colonial environmental state and for considering how ongoing dispossessions of lands and resources are finding analogous expression in what Leanne Simpson calls "digital dispossession." See Coulthard and Simpson, "Grounded Normativity."

13 Simpson, *As We Have Always Done.*

14 Fairbanks et al., "Assembling Enclosure."

15 SQCRD, "Regular Board Meeting Held at 344 2nd Avenue West," 55.

16 When I first came upon the detail—buried in the notes of a regional update document—I was engrossed with my research on Dodge Cove's noise abatement campaign, while the community was engrossed with fighting a new industrial project—the proposed Nexxen LNG facility—that had lately usurped their concerns with the port. It would be another ten months before I was able to have a detailed conversation about the shore station with any community members.

17 See ONC, "Prince Rupert—Ts'msyen Territory Community Observatory."

18 ONC, "Prince Rupert—Ts'msyen Territory Community Observatory."

19 Another Smart Oceans linkage to oil and gas comes through Transport Canada, which oversees shipping and has developed a Smart Oceans Contribution Program with ONC. The quote is from Transport Canada, "Ocean Networks Canada."

20 See Moore, "B.C. Ocean Observation Project Useful for Oil Industry."

21 Chun, *Updating to Remain the Same*, 27.

22 The Canada Foundation for Innovation (CFI), Western Economic Diversification Canada, and Transport Canada.

23 Halpern and Mitchell, *The Smartness Mandate*, 9.

24 Helmreich, "Wave Theory ~ Social Theory," 315.

25 In agreeing to host Smart Oceans infrastructure, different North Coast communities would hold detailed negotiations with ONC representatives. I was not privy to these conversations and do not make the assumption that all featured the same final agreements. In Hartley Bay, for instance, I was made aware that Gitga'at decision-makers insisted on strict protections for all hydrophone data recorded in Gitga'at Territory, which might not have been the case in other instances.

26 ONC, "Products and Services Deliverables Report," 000203.

27 ONC, "Teacher Info," 2.

28 Today the course is called ACE 196 Instrument Technology and is offered by the newly branded Coast Mountain College (accessed November 9, 2018, https://catalogue.coastmountaincollege.ca/courses/ace196/).

29 Maguire and Winthereik, "Digitalizing the State." For the development policy, see Government of Canada, "The Prime Minister of Canada Announces the National Oceans Protection Plan."

30 Adams and Matsumodo, "Enhancing Ocean Literacy Using Real-Time Data," 9.
31 See ONC, "Products and Services Deliverables Report," 000198.
32 ONC, "Teacher Info," 5.
33 ONC, "Indigenous Community Engagement Plan." Obtained under Government of Canada Access to Information (ATI) request A 2017–01974. Another salient example here is ONC's Youth Science Ambassadors—a project that engages students to "respectfully connect" Indigenous knowledge to ONC platforms.
34 ONC, "Teacher Info," 7.
35 Sevilla-Buitrago, "Capitalist Formations of Enclosure." The idea chimes with the insights of various internet critics who have pointed out how the technology's seemingly unbounded and free-floating innovations depended historically on public funding, infrastructure, and unwaged work. For example, see Tarnoff, "The Internet Should be a Public Good."
36 See, for example, Galloway and Thacker, *The Exploit*; Chun, *Updating to Remain the Same.*
37 Coulthard, *Red Skins, White Masks*, 15.
38 It was later termed the Indigenous Community Engagement Plan. See ONC, "Indigenous Community Engagement Plan."
39 ONC, "Strategic Plan 2030," 21. See also ONC, "Indigenous Community Engagement Plan."
40 ONC, "Strategic Plan 2030," 21. In the intervening years, the organization had hired a First Nations Engagement Coordinator and commenced dialogues with several nations (see ONC, "Indigenous Community Engagement Plan," 62–97).
41 See my paper coauthored with Mike Simpson: Ritts and Simpson, "Smart Oceans Governance."
42 See Norgard et al., "Northeast Pacific Seamount Expedition," 42–43. This work was instrumental in establishing the management plan for the site. See Government of Canada, "SGaan Kinghlas–Bowie Seamount."
43 Byrd, "'Do They Not Have Rational Souls?,'" 433.
44 Kwan, "Algorithmic Geographies," 274. While I have spoken with marine planners with the Council of the Haida Nation about this topic, my perspective is more informed by Hartley Bay, where the Gitga'at Nation has made several recent hires to resolve challenges around the size of incoming data about the territory. But I do not claim to represent either Nation here.
45 The episode (and others like it) is explored in Simlai, "Negotiating the Panoptic Gaze."
46 Heesemann et al., "Ocean Networks Canada," 153. This includes $8.9 million from the provincial government in 2012 to support ongoing research activities; $20 million from Transport Canada in 2014 to aid government efforts to achieve a World-Class Tanker Safety System; infrastructure funding through the Canadian Foundation for Innovation; and personnel funding through the quasi-science agencies NSERC and Polar Knowledge Canada. For more on the

funding provisions from the Canadian government, see Ritts and Simpson, "Smart Oceans Governance."

47 This dynamic should give us pause when confronted with critical valorizations of "becoming sensor." Myers, "Becoming Sensor in Sentient Worlds."

48 Halpern and Mitchell, *The Smartness Mandate*, 7.

49 There is some evidence to suggest that ONC itself was struggling to deal with this problem. By 2017, it had over 258 terabytes of regionally sourced sound under its management. In 2020, an internal estimate put this number at over seven hundred total. By this time, there were almost five thousand ONC sensors across Canadian waters reporting data—the vast majority on the BC coast. The Great Bear Sea Network had over 55,000 hours of recordings to sift through; Cetacea Lab had processed three thousand hours of hydrophone activity and had months more systematizing to do.

50 Heesemann et al., "Ocean Networks Canada." Ocean Networks Canada uses CANARIE, a data server system located in a bunker in central Canada, to store its data. CANARIE links all Canadian universities so that they can share data.

51 Ocean Networks Canada, Device Details, Oceans 2.0 (note that this is now Oceans 3.0), https://data.oceannetworks.ca/DeviceListing?DeviceId=23482.

52 Bruyninckx, *Listening in the Field*, 160.

53 ONC, "March Report," 000046.

54 Kahn, "Concerning the Line," 180.

55 Much like the locations portrayed in Google Maps, spectral probability density maps are stitched together from different kinds of data protocols. They are not a synthesis of sonic space so much as the assembling of discrete points to model an appearance as such. Meanwhile, the unwired spaces out in the actual location are unavailable here, while certain sounds, like those made by the Canadian Navy, are made to be unavailable—scrubbed from the resulting plot. As such, the spectral probability density map complicates the indicative promise of a Smart Ocean, revealing machine listening less as a fully fledged agency than a computationally intensive form of pattern recognition, unevenly informed by human bias and evaluation datasets. The navy periodically requests to have sensitive recording frequencies filtered out because recordings may contain signals bearing military activity. When filtering occurs, the filename is appended with HPF or LPF for high-pass or low-pass filtering. Sometimes, the military-diverted data will be returned at a later date, but modified from the original data. See Braga, "Listening In."

56 The estimate is given by David Hannay in Gaetz, "Company Responsible for Underwater Noise Monitoring Project."

57 Department of Fisheries and Oceans, "Management Measures." See related work connected to ECHO on the Department of Fisheries website, accessed March 6, 2024, https://www.pac.dfo-mpo.gc.ca/fm-gp/mammals-mammiferes/whales-baleines/srkw-measures-mesures-ers-eng.html.

58 DFO, "Management Measures."

59 See Government of Canada, "The Prime Minister of Canada Announces the National Oceans Protection Plan."
60 ONC, "Products and Services Deliverables Report," 000198.
61 ONC, "Products and Services Deliverables Report," 000232.
62 Virilio, *The Original Accident*. Recognizing the dependencies and fragilities of their systems, the federal and provincial governments have commenced new efforts to upgrade the coast's internet infrastructure and associated power grids. The stresses can only be expected to increase, however, as demands for integrated systems move to encompass outer sections of the coast still limited in terms of wi-fi and boat access.
63 Hunter, "The Sinking of the Nathan E. Stewart."
64 Hunter, "The Sinking of the Nathan E. Stewart."
65 Turkle, *Alone Together*, 13. Turkle's conceptualization of "weak ties" helps to explain how a network-driven sociality might reduce agency to procedural effects and social relationships to cross-functional transactions.
66 Luque-Ayala and Marvin, *Urban Operating Systems*, 224.
67 Jameson, *Archaeologies of the Future*.
68 Helmreich, *Alien Ocean*, 243.

CONCLUSION

1 Calami, "Was Chief's Chant 'Hey, Hey' or 'Ha, Ha'?"
2 The characteristic refusal in Clifton's gesture, with its broad recognizability but also deeply felt local sensibility, was plainly audible to Jeremy and Spencer, who laughed for a good while when I shared the article with them.
3 Campling and Colás, *Capitalism and the Sea*, 203.
4 Country music is a powerfully generative field for Indigeneity, celebrated by Indigenous peoples throughout Native North America, from Navajo Country to Dene Lands to the territories of the Mikmaq to Hartley Bay. For work on country music and Indigeneity, see Samuels, *Putting a Song on Top of It*; Jacobsen, *The Sound of Navajo Country*; and Shuvera, "Southern Sounds, Northern Voices."
5 Hubbs, "Is Country Music Quintessentially American?"
6 Gopinath and Schultz, "Sentimental Remembrance and the Amusements of Forgetting," 481.
7 Mann, "Why Does Country Music Sound White?"
8 Jacobsen, *The Sound of Navajo Country*, 15.
9 Adorno, "Music in the Background."
10 There's a historical element to this, many of Prince Rupert's now-vacated spaces of collective gathering—from Dillon's Country Bar (later Johnny Bs), the Savoy, the Royal, the Surf Club—having also been spaces of country music.
11 I coined this phrase during the tour with Jeremy Pahl, Jess Rampling, and Simone Schmidt.

12 On belonging in country music, see especially Hubbs, "Is Country Music Quintessentially American?"

13 Charland, "Technological Nationalism," 197. This approach would also be able to examine the nationalistic currents running through several of the sound cultures referenced in this book, such as acoustic ecology or, indeed, country music (see Schafer, *The Book of Noise*; Schafer, *The Tuning of the World*). It would also be an occasion to reconsider claims made in a chapter in Jody Berland's *North of Empire* titled "Locating Listening," a wonderful, if somewhat sweeping, formulation of radio programming and broadcasting as a tool of Canadian nation building. More work could be done to survey the specific imprint of national radio and liberalism in the audiopolitics of the North Coast. For some useful insights to this end, see Atkinson, "Replacing Sound Assumptions."

14 Peters, *Speaking into the Air*, 161.

15 I am thinking, for example, of the emerging work of Diego Ellis Soto. See his website at https://diegoellissoto.org. In particular, consider the lecture titled "Collective Pulse," also available on YouTube: "Collective Pulse: Uncovering Hidden Lives of Animals through Music Theory & Artificial Intelligence," posted by Yale University, https://www.youtube.com/watch?v=nYN5RS6r2ZE&t=3s.

16 Lacey, *Listening Publics*, 197.

17 Randon, Dowd, and Joy, "A Real-Time Data Assimilative Forecasting System."

18 Bowker, *Memory Practices in the Sciences*, 120.

19 Sugai and Llusia, "Bioacoustic Time Capsules."

20 Keen et al., "Ship-Strike Forecast and Mitigation."

21 Cecco, "How a Huge New LNG Plant Spells 'Dire' Trouble for Whales off Canada's Coast."

22 Jasper, "Sonic Refugia."

23 Collaborators also include the World Wildlife Fund, SIMRES, OrcaLab, and BC Whales. See Whale Sound, "Who We Are," https://whalesound.ca/who-we-are/.

24 Sterne describes the "sonic imagination" in terms of "a figural practice that reaches into fields of sonic knowledge and practice, and blends them with other questions, problems, fields, spaces and histories." See Sterne, "Sonic Imaginations," 6.

25 Ekers and Loftus, "Revitalizing the Production of Nature Thesis," 237.

26 Haraway, "Situated Knowledges."

27 Gray, "Rematriation," 3. Institutionally engaged rematriation efforts, led by Gray, were taking shape as this book went to print. See https://robingray.ca.

28 Denning, *Noise Uprising*, 10. I venture this claim mindful of my own limited perspective on the topic. As Gray writes, "Non-Indigenous people . . . likely will not fully account for what rematriation is, what it does, what it wants, and what it takes." See Gray, "Rematriation," 5.

BIBLIOGRAPHY

Adams, Lisa, and George Matsumoto. "Enhancing Ocean Literacy Using Real-Time Data." *Oceanography* 22, no. 2 (2009): 8–9. https://doi.org/10.5670/oceanog.2009.55.

Adorno, Theodor. *Essays on Music.* Selected by Richard Leppert, translated by Susan H. Gillespie. Berkeley: University of California Press, 2002

Adorno, Theodor. "Late Style in Beethoven." 1937. In *Essays on Music,* selected by Richard Leppert, translated by Susan H. Gillespie, 564–68. Berkeley: University of California Press, 2002.

Adorno, Theodor. "Music in the Background." Circa 1934. In *Essays on Music,* selected by Richard Leppert, translated by Susan H. Gillespie, 506–512. Berkeley: University of California Press, 2002.

Adorno, Theodor. "On the Fetish-Character of Music and the Regression of Listening." 1938. In *Essays on Music,* selected by Richard Leppert, translated by Susan H. Gillespie, 288–318. Berkeley: University of California Press, 2002.

Adorno, Theodor. *Philosophy of Modern Music.* Translated by Anne G. Mitchell and Wesley V. Blomster. New York: Continuum, 1994.

Affolderbach, Julia, Roger Alex Clapp, and Roger Hayter. "Environmental Bargaining and Boundary Organizations: Remapping British Columbia's Great Bear Rainforest." *Annals of the Association of American Geographers* 102, no. 6 (2012): 1391–408. http://dx.doi.org/10.1080/00045608.2012.706567.

Agardy, Tundi, Giuseppe Notarbartolo di Sciara, and Patrick Christie. "Mind the Gap: Addressing the Shortcomings of Marine Protected Areas through Large Scale Marine Spatial Planning." *Marine Policy* 35, no. 2 (2011): 226–32. https://doi.org/10.1016/j.marpol.2010.10.006.

Ainslie, M. *Principles of Sonar Performance Modeling.* New York: Springer, 2010.

Akhter, Majed. "Infrastructure Nation: State Space, Hegemony, and Hydraulic Regionalism in Pakistan." *Antipode* 47, no. 4 (2015): 849–70.

Alger, Justin, Jane Lister, and Peter Dauvergne. "Corporate Governance and the Environmental Politics of Shipping." *Global Governance: A Review of Multilateralism and International Organizations* 27, no. 1 (2021): 144–66.

Allison, Lou, and Jane Wilde, eds. *Gumboot Girls: Adventure, Love and Survival on the North Coast of British Columbia*. Qualicum Beach, BC: Caitlin, 2014.

Althusser, Louis. *On the Reproduction of Capitalism: Ideology and Ideological State Apparatuses*. New York: Verso, 2014.

Ambach, Mike. "Current and Proposed Projects with Underwater Noise Implications for the North and Central Coast." World Wildlife Fund, 2013.

Ames, Kenneth M. *The North Coast Prehistory Project Excavations in Prince Rupert Harbour, British Columbia: The Artifacts*. Oxford: British Archaeological Reports, 2005.

Andrejevic, Mark. "The Pacification of Interactivity." In *The Participatory Condition in the Digital Age*, 187–206. Minneapolis: University of Minnesota Press, 2016. https://research.monash.edu/en/publications/the-pacification-of-interactivity.

Andrejevic, Mark. "Surveillance in the Digital Enclosure." In *The New Media of Surveillance*, edited by Shoshana Magnet and Kelly Gates, 18–40. New York: Routledge, 2009.

Andrejevic, Mark. "Ubiquitous Computing and the Digital Enclosure Movement." *Media International Australia* 125, no. 1 (2007): 106–17. https://doi.org/10.1177/1329878X0712500112.

Archer, David. "Village Patterns and the Emergence of Ranked Society in the Prince Rupert Area." In *Perspectives on Northern Northwest Coast Prehistory*, edited by Jerome S. Cybulski, 203–22. Mercury Series. Ottawa: University of Ottawa Press, 2001.

Artelle, K. A., M. S. Adams, H. M. Bryan, C. T. Darimont, J. ('Cúagilákv) Housty, W. G. (Dúqváísḷa) Housty, J. E. Moody, et al. "Decolonial Model of Environmental Management and Conservation: Insights from Indigenous-Led Grizzly Bear Stewardship in the Great Bear Rainforest." *Ethics, Policy and Environment* 24, no. 3 (2021): 283–323. https://doi.org/10.1080/21550085.2021.2002624.

Atkinson, Maureen. "Replacing Sound Assumptions: Rediscovered Narratives of Post War Northern British Columbia." PhD diss., University of Waterloo, 2017. https://uwspace.uwaterloo.ca/handle/10012/12571.

Attali, Jacques. *Noise: The Political Economy of Music*. Translated by Brian Massumi. Manchester, UK: Manchester University Press, 1985.

Bakhtin, Mikhail. *The Dialogic Imagination*. Edited by M. Holquist, translated by C. Emerson and M. Holquist. Austin: University of Texas Press, 1981.

Bakker, Karen. *The Sounds of Life: How Digital Technology Is Bringing Us Closer to the Worlds of Animals and Plants*. Princeton, NJ: Princeton University Press, 2022.

Bakker, Karen, and Max Ritts. "Smart Earth: A Meta-review and Implications for Environmental Governance." *Global Environmental Change* 52 (September 2018): 201–11.

Barbeau, Marius. *Pathfinders in the North Pacific*. Caldwell, ID: Caxton Printers, 1958.

Barbeau, Marius. "Songs of the Northwest." *Musical Quarterly* 19, no. 1 (1933): 101–11.

Barbeau, Marias, and William Beynon. *Tsimshian Narratives*. Vol. 1, *Tricksters, Shamans, and Heroes*. Ottawa: University of Ottawa Press, 1987. https://doi.org/10.2307/j.ctv171b4.

Barman, Jean. "The West beyond the West: The Demography of Settlement in British Columbia." *Journal of Canadian Studies* 25, no. 3 (1990): 5–18. https://doi.org/10.3138/jcs.25.3.5.

Barua, Maan. "Animating Capital: Work, Commodities, Circulation." *Progress in Human Geography* 43, no. 4 (2019): 650–69. https://doi.org/10.1177/0309132518819057.

Bateson, Gregory. "The Logical Categories of Learning and Communication." In *Steps to an Ecology of Mind: Collected Essays in Anthropology, Psychiatry, Evolution, and Epistemology*, 279–308. 1972. Chicago: University of Chicago Press, 2000.

Battistoni, Alyssa. "Bringing in the Work of Nature: From Natural Capital to Hybrid Labor." *Political Theory* 45, no. 1 (2017): 5–31. https://doi.org/10.1177/0090591716638389.

Baumgartner, Mark F., Julianne Bonnell, Sofie M. Van Parijs, Peter J. Corkeron, Cara Hotchkin, Keenan Ball, Léo-Paul Pelletier, et al. "Persistent Near Real-Time Passive Acoustic Monitoring for Baleen Whales from a Moored Buoy: System Description and Evaluation." *Methods in Ecology and Evolution* 10, no. 9 (2019): 1476–89. https://doi.org/10.1111/2041-210X.13244.

Benjamin, Walter. *Illuminations*. Translated by Harry Zohn. New York: Schocken Books, 1968.

Berger, Harris M. *Metal, Rock, and Jazz: Perception and the Phenomenology of Musical Experience*. Middletown, CT: Wesleyan University Press, 1999.

Berland, Jody. "Locating Listening." In *North of Empire: Essays on the Cultural Technologies of Space*, 185–209. Durham, NC: Duke University Press, 2009.

Berland, Jody. *North of Empire: Essays on the Cultural Technologies of Space*. Durham, NC: Duke University Press, 2009. https://muse.jhu.edu/pub/4/monograph/book/69131.

Besky, Sarah, and Alex Blanchette, eds. *How Nature Works: Rethinking Labor on a Troubled Planet*. Albuquerque: University of New Mexico Press, 2019.

Bijsterveld, Karin. *Sonic Skills: Listening for Knowledge in Science, Medicine and Engineering (1920s–Present)*. Berlin: Springer Nature, 2019.

Birtchnell, Thomas, Satya Savitzky, and John Urry, eds. *Cargomobilities: Moving Materials in a Global Age*. Abingdon, UK: Routledge, 2019.

Blanchet, M. Wylie. *The Curve of Time*. Potomac, MD: Pickle Partners, 2016.

Boas, Franz. "On Alternating Sounds." *American Anthropologist* 2, no. 1 (1889): 47–54.

Born, Georgina. "On Musical Mediation: Ontology, Technology and Creativity." *Twentieth-Century Music* 2, no. 1 (2005): 7–36. https://doi.org/10.1017/S147857220500023X.

Bosworth, Kai. *Pipeline Populism: Grassroots Environmentalism in the Twenty-First Century*. Minneapolis: University of Minnesota Press, 2022.
Bowker, Geoffrey C. *Memory Practices in the Sciences*. Cambridge, MA: MIT Press, 2008.
Bowles, Paul, and Fiona MacPhail. "The Town That Said 'No' to the Enbridge Northern Gateway Pipeline: The Kitimat Plebiscite of 2014." *Extractive Industries and Society* 4, no. 1 (2017): 15–23. https://doi.org/10.1016/j.exis.2016.11.009.
Bowman, Phyllis. "Rail Operations Declined after Waterfront Fire." *Prince Rupert Daily News*, June 12, 1997.
Braga, Matthew. "Listening In: The Navy Is Tracking Ocean Sounds Collected by Scientists." *Atlantic*, August 18, 2014. https://www.theatlantic.com/technology/archive/2014/08/listening-in-the-navy-is-tracking-ocean-sounds-collected-by-scientists/378630/.
Bratton, Benjamin H. *The Stack: On Software and Sovereignty*. Cambridge, MA: MIT Press, 2016.
Braun, Bruce. *The Intemperate Rainforest: Nature, Culture, and Power on Canada's West Coast*. Minneapolis: University of Minnesota Press, 2002.
Bray, Cathy. "Watching the Whale Watchers: Leisurely Informal Learning Online." OrcaLive, 2004. Accessed April 20, 2016. http://www.orca-live.net.
Bridge, Gavin, and Tom Perreault. "Environmental Governance." In *A Companion to Environmental Geography*, edited by Noel Castree, David Demeritt, Diana Liverman, and Bruce Rhoads, 475–97. Hoboken, NJ: John Wiley, 2009. https://doi.org/10.1002/9781444305722.ch28.
Brown, Wendy. *States of Injury: Power and Freedom in Late Modernity*. Princeton, NJ: Princeton University Press, 2020.
Bruyninckx, Joeri. *Listening in the Field: Recording and the Science of Birdsong*. Cambridge, MA: MIT Press, 2018.
Budd, Robert. "'The Story of the Country': Imbert Orchard's Quest for Frontier Folk in BC, 1870–1914." MA thesis, University of Victoria, 2005. http://dspace.library.uvic.ca/bitstream/handle/1828/856/budd_2005.pdf.
Bull, Michael. "No Dead Air! The iPod and the Culture of Mobile Listening." *Leisure Studies* 24, no. 4 (2005): 343–55. https://doi.org/10.1080/0261436052000330447.
Bürkner, Hans-Joachim, and Bastian Lange. "Sonic Capital and Independent Urban Music Production: Analysing Value Creation and 'Trial and Error' in the Digital Age." *City, Culture and Society* 10 (2017): 33–40. https://doi.org/10.1016/j.ccs.2017.04.002.
Burnett, D. Graham. *The Sounding of the Whale: Science and Cetaceans in the Twentieth Century*. Chicago: University of Chicago Press, 2013.
Büscher, Bram. "Nature 2.0: Exploring and Theorizing the Links between New Media and Nature Conservation." *New Media and Society* 18, no. 5 (2016): 726–43. https://doi.org/10.1177/1461444814545841.
Büscher, Bram, and Robert Fletcher. *The Conservation Revolution: Radical Ideas for Saving Nature beyond the Anthropocene*. New York: Verso, 2020.

Byrd, Jodi A. "'Do They Not Have Rational Souls?': Consolidation and Sovereignty in Digital New Worlds." *Settler Colonial Studies* 6, no. 4 (2016): 423–37. https://doi.org/10.1080/2201473X.2015.1090635.

Byrd, Jodi A. "Follow the Typical Signs: Settler Sovereignty and Its Discontents." *Settler Colonial Studies* 4, no. 2 (2014): 151–54. https://doi.org/10.1080/2201473X.2013.846388.

Byrd, Jodi A. *The Transit of Empire: Indigenous Critiques of Colonialism*. Minneapolis: University of Minnesota Press, 2011.

Calami, Peter. "Was Chief's Chant 'Hey, Hey' or 'Ha, Ha'?" *Native Voice*, May 1975. File GR-1738.110.25—Native Voice. Royal BC Museum, Victoria, BC.

Campbell, Kevin. "Friction Grows over Dodge Cove Community Plan—Prince Rupert Northern View." *Northern View*, March 8, 2017, sec. News. https://www.thenorthernview.com/news/friction-grows-over-dodge-cove-community-plan/.

Campling, Liam, and Alejandro Colás. *Capitalism and the Sea: The Maritime Factor in the Making of the Modern World*. New York: Verso, 2021.

Camprubí, Lino, and Alexandra Hui. "Testing the Underwater Ear." In *Testing Hearing: The Making of Modern Aurality*, edited by Viktoria Tkaczyk, Mara Mills, and Alexandra Hui, 301–26. Oxford: Oxford University Press, 2020.

Carlson, Thomas J., Christ de Jong, and Rene Dekeling. "Workshop One: Risk Analysis." In *The Effects of Noise on Aquatic Life*, edited by Arthur N. Popper and Anthony Hawkins, 657–59. London: Springer, 2012.

CBC. "Saltwater Brothers." Daybreak North, March 18, 2013. Accessed April 20, 2016. http://www.cbc.ca/daybreaknorth/interviews/2013/03/18/salt-water-brothers-extended-interview/.

Cecco, Leyland. "How a Huge New LNG Plant Spells 'Dire' Trouble for Whales off Canada's Coast." *Guardian*, October 7, 2023. https://www.theguardian.com/environment/2023/oct/07/how-a-huge-new-lng-plant-spells-dire-trouble-for-whales-off-canadas-coast.

Chapin, Tom, and John Forster. *Sing a Whale Song*. New York: Random House, 1993.

Chapman, Ross, and Andrea Price. "Low Frequency Deep Ocean Ambient Noise Trend in the Northeast Pacific Ocean." *Journal of the Acoustical Society of America* 129, no. 5 (2011). https://doi.org/10.1121/1.3567084.

Charland, Maurice. "Technological Nationalism." *Canadian Journal of Political and Social Theory* 10, no. 1–2 (1986): 196–220.

Charnock, Greig. "Challenging New State Spatialities: The Open Marxism of Henri Lefebvre." *Antipode* 42, no. 5 (2010): 1279–303.

Chion, Michel. *Music in Cinema*. Edited and translated by Claudia Gorbman. New York: Columbia University Press, 2021.

Chua, Charmaine. "Containing the Ship of State: Managing Mobility in an Age of Logistics." PhD diss., University of Minnesota, 2018. http://conservancy.umn.edu/handle/11299/200214.

Chun, Wendy Hui Kyong. *Updating to Remain the Same: Habitual New Media*. Cambridge, MA: MIT Press, 2016.

Citizen Staff. "Are Prince George White Supremacists Gaining Momentum?" *Prince George Citizen*, February 12, 2011. https://www.princegeorgecitizen.com/local-news/are-prince-george-white-supremacists-gaining-momentum-3700302.

Clark, Christopher, William Ellison, Brandon L. Southall, Leila Hatch, Sofie Van Parijs, Adam Frankel, and Dimitri Ponirakis. "Acoustic Masking in Marine Ecosystems: Intuitions, Analysis, and Implication." *Marine Ecology Progress Series* 395 (2009): 201–22. https://doi.org/10.3354/meps08402.

Coastal First Nations. "Eyes and Ears of the Land and Sea." July 18, 2022. https://coastalfirstnations.ca/eyes-and-ears-of-the-land-and-sea/.

Coast Funds. "From Conflict to Collaboration." Accessed July 17, 2023. https://coastfunds.ca/resources/conflict-to-collaboration/.

Colby, Jason M. *Orca: How We Came to Know and Love the Ocean's Greatest Predator*. New York: Oxford University Press, 2018.

Cole, Douglas. *Captured Heritage: The Scramble for Northwest Coast Artifacts*. Vancouver: UBC Press, 1995.

Collard, Rosemary-Claire. *Animal Traffic: Lively Capital in the Global Exotic Pet Trade*. Durham, NC: Duke University Press, 2020.

Connor, Steven. "The Menagerie of the Senses." *The Senses and Society* 1, no. 1 (2006): 9–26.

Corbin, Alain. *Village Bells: The Culture of the Senses in the Nineteenth-Century French Countryside*. Translated by Martin Thom. New York: Columbia University Press, 1998.

Coulthard, Glen Sean. *Red Skin, White Masks: Rejecting the Colonial Politics of Recognition*. Minneapolis: University of Minnesota Press, 2014.

Coulthard, Glen, and Leanne Betasamosake Simpson. "Grounded Normativity/Place-Based Solidarity." *American Quarterly* 68, no. 2 (2016): 249–55.

Coupland, G., K. Stewart, and T. J. Hall. *Archaeological Investigations in the Prince Rupert Area in 1999 and 2000: The Differentiated Regional Economy at Prehistoric Prince Rupert. Permit 1999–138*. Victoria, BC: British Columbia Archaeology Branch, 2002.

Cox, Christoph. *Sonic Flux: Sound, Art, and Metaphysics*. Chicago: University of Chicago Press, 2018.

Crawford, Kate, and Vladan Joler. "Anatomy of an AI System: The Amazon Echo as an Anatomical Map of Human Labor, Data and Planetary Resources." AI Now Institute and Share Lab, September 7, 2018. http://www.anatomyof.ai.

Cruikshank, Julie. *Do Glaciers Listen? Local Knowledge, Colonial Encounters, and Social Imagination*. Vancouver: UBC Press, 2007.

Crumb, George. *Vox Balaenae*. 1971. Naxos. 8.559205.

Cusick, Suzanne G. "Towards an Acoustemology of Detention in the 'Global War on Terror.'" In *Music, Sound and Space: Transformations of Public and Private Experience*, edited by Georgina Born, 275–91. Cambridge: Cambridge University Press, 2013. https://doi.org/10.1017/CBO9780511675850.017.

Cvetkovich, Ann. *Depression: A Public Feeling*. Durham, NC: Duke University Press, 2012.

Cybulski, Jerome. "Village Patterns and the Emergence of Ranked Society in the Prince Rupert Area." In *Perspectives on Northern Northwest Coast Prehistory*, edited by Jerome S. Cybulski, 203–22. Mercury Series. Ottawa: University of Ottawa Press, 2001.

Daloz, Kate. *We Are as Gods: Back to the Land in the 1970s on the Quest for a New America*. New York: PublicAffairs, 2016.

Da Luz, Nuno. "Undocumented Birds: Echopolitics as Interspecies Resistance." *Antropoceno Sónico* 10, no. 1 (2021): 26–36.

Daughtry, J. Martin. "Thanatosonics: Ontologies of Acoustic Violence." *Social Text* 32, no. 2 (2014): 25–51. https://doi.org/10.1215/01642472-2419546.

David, Léa. *Progress Report on the Implementation of an ACCOBAMS Certification for Highly Qualified MMOs/PAM*. Thirteenth Meeting of the ACCOBAMS Scientific Committee, February 26–28, 2020. Accessed April 18, 2021. https://accobams.org/wp-content/uploads/2020/01/SC13.Inf13_Progress-Report-on-implementation-of-an-ACCOBAMS-certification.pdf.

Dawkins, Ashley, and Alex Loftus. "The Senses as Direct Theoreticians in Practice." *Transactions of the Institute of British Geographers* 38, no. 4 (2013): 665–77. https://doi.org/10.1111/j.1475-5661.2012.00551.x.

Debord, Guy. *Society of the Spectacle*. Detroit: Black and Red, 2002.

DeCola, Elise, Tim Robertson, Andrew Mattox, David Eley, and Sierra Fletcher. *Nuka Research and Planning Group West Coast Oil Response Study*. Vol. 2, *Vessel Traffic Study*. Plymouth, MA: Nuka Research and Planning Group, 2013. https://www2.gov.bc.ca/assets/gov/environment/air-land-water/spills-and-environmental-emergencies/docs/westcoastspillresponse_vol2_vesseltrafficstudy_130722.pdf.

De Leeuw, Sarah, Emilie S. Cameron, and Margo L. Greenwood. "Participatory and Community-Based Research, Indigenous Geographies, and the Spaces of Friendship: A Critical Engagement." *Canadian Geographer* 56, no. 2 (2012): 180–94. https://doi.org/10.1111/j.1541-0064.2012.00434.x.

Dembicki, Geoff. "At Ground Zero for Next Huge Enviro War: Hartley Bay." *Tyee*, September 13, 2010. https://thetyee.ca/News/2010/09/13/HartleyBayEnviroWar/.

Dempsey, Jessica. "Biodiversity Loss as Material Risk: Tracking the Changing Meanings and Materialities of Biodiversity Conservation." *Geoforum* 45 (2013): 41–51. http://dx.doi.org/10.1016/j.geoforum.2012.04.002.

Dempsey, Jessica. "The *Politics of Nature* in British Columbia's Great Bear Rainforest." *Geoforum* 42, no. 2 (2011): 211–21.

Denning, Michael. *Noise Uprising: The Audiopolitics of a World Musical Revolution*. New York: Verso, 2015.

Department of Fisheries and Oceans (DFO). "Species Listing under Canada's Species at Risk Act." *Conservation Biology* 23, no. 6 (2009): 1609–17.

Department of Fisheries and Oceans (DFO). "Management Measures to Protect Southern Resident Killer Whales." Department of Fisheries and Oceans. Accessed September 18, 2022. https://www.pac.dfo-mpo.gc.ca/fm-gp/mammals-mammiferes/whales-baleines/srkw-measures-mesures-ers-eng.html.

Dobell, Darcy. "Caamaño: The Sound of (Whale) Music." *Hakai Magazine*, January 29, 2016. https://hakaimagazine.com/article-short/caamano-sound-whale-music/.

Dolar, Mladen. *A Voice and Nothing More*. Cambridge, MA: MIT Press, 2006.

Dow, M. T., Emling, J. W., Knudsen V. "Survey of Underwater Sounds, Sounds from Surface Ships." 1945. O.S. R. D., Div. 6.1. N.D.R.C. Rept. (Declassified 1960).

Drott, Eric. *Streaming Music, Streaming Capital*. Durham, NC: Duke University Press, 2024.

Duarte, Carlos M., Lucille Chapuis, Shaun P. Collin, Daniel P. Costa, Reny P. Devassy, Victor M. Eguiluz, Christine Erbe, et al. "The Soundscape of the Anthropocene Ocean." *Science* 371 (2021): eaba4658. https://doi.org/10.1126/science.aba4658.

Duffus, David A., and Philip Dearden. "Non-consumptive Wildlife-Oriented Recreation: A Conceptual Framework." *Biological Conservation* 53, no. 3 (1990): 213–31. https://doi.org/10.1016/0006-3207(90)90087-6.

Edwards, Paul N. *A Vast Machine: Computer Models, Climate Data, and the Politics of Global Warming*. Cambridge, MA: MIT Press, 2013.

Ekers, Mike, and Alex Loftus. "Revitalizing the Production of Nature Thesis: A Gramscian Turn?" *Progress in Human Geography* 37, no. 2 (2012): 234–52. https://doi.org/10.1177/0309132512448831.

Enbridge. "Northern Gateway Application ('App.') Vol. 8B: Marine Transportation es." 2010. http://www.dfo-mpo.gc.ca/csas-sccs/Publications/ScR-RS/2012/2012_034-eng.pdf.

Environmental Assessment Office (EAO). "Collected Public Comments for Aurora LNG Digby Island Project September 1, 2015 to October 1, 2015." 2015. Accessed September 1, 2017. https://projects.eao.gov.bc.ca.

Erbe, C., A. Duncan, and M. Koessler. "Modelling Noise Exposure Statistics from Current and Projected Shipping Activity in Northern British Columbia." 2012. Perth: Centre for Marine Science and Technology, Curtin University.

Erbe, Christine, and David M. Farmer. "A Software Model to Estimate Zones of Impact on Marine Mammals around Anthropogenic Noise." *Journal of the Acoustical Society of America* 108, no. 3 (2000): 1327–31. https://doi.org/10.1121/1.1288939.

Erbe, Christine, Alexander MacGillivray, and Rob Williams. "Mapping Cumulative Noise from Shipping to Inform Marine Spatial Planning." *Journal of the Acoustical Society of America* 132, no. 5 (2012): EL423–28. https://doi.org/10.1121/1.4758779.

Erbe, Christine, Rob Williams, Doug Sandilands, and Erin Ashe. "Identifying Modeled Ship Noise Hotspots for Marine Mammals of Canada's Pacific Region." *PLOS ONE* 9, no. 3 (2013): e89820. https://doi.org/10.1371/journal.pone.0089820.

European Commission. "Marine Environment: EU Policies to Protect Europe's Oceans, Seas and Coasts." Accessed July 29, 2023. https://environment.ec.europa.eu/topics/marine-and-coastal-environment_en.

Fairbanks, Luke, Lisa M. Campbell, Noëlle Boucquey, and Kevin St. Martin. "Assembling Enclosure: Reading Marine Spatial Planning for Alternatives." *Annals of the American Association of Geographers* 108, no. 1 (2018): 144–61. https://doi.org/10.1080/24694452.2017.1345611.
Fanon, Frantz. *The Wretched of the Earth.* Translated by Richard Philcox. New York: Grove Press, 1963.
Farmer, David. "Acoustical Studies of the Upper Ocean Boundary Layer." In *Sounds in the Sea: From Ocean Acoustics to Acoustical Oceanography*, edited by Herman Medwin, 315–40. Cambridge: Cambridge University Press, 2005.
Farmer, David M., Svein Vagle, and A. Donald Booth. "A Free-Flooding Acoustical Resonator for Measurement of Bubble Size Distributions." *Journal of Atmospheric and Oceanic Technology* 15, no. 5 (1998): 1132–46.
Favali, Paolo, Laura Beranzoli, and Angelo De Santis. *Seafloor Observatories: A New Vision of the Earth from the Abyss.* Berlin: Springer Science and Business Media.
Fee, Margery. "Rewriting Anthropology and Identifications on the North Pacific Coast: The Work of George Hunt, William Beynon, Franz Boas, and Marius Barbeau." *Australian Literary Studies* 25, no. 4 (2010): 17–32.
Fletcher, Robert. "Gaming Conservation: Nature 2.0 Confronts Nature-Deficit Disorder." *Geoforum* 79 (2017): 153–62. https://doi.org/10.1016/j.geoforum.2016.02.009.
Fletcher, Robert. "Taking the Chocolate Laxative: Why Neoliberal Conservation 'Fails Forward.'" In *Nature Inc.: Environmental Conservation in the Neoliberal Age*, edited by Bram Büscher, Wolfram Dressler, and Robert Fletcher, 87–107. Tucson: University of Arizona Press, 2014.
Ford, John, Robert Abernethy, John Phillips, George Calambokidis, Graeme Ellis, and Linda M. Nichol. *Distribution and Relative Abundance of Cetaceans in Western Canadian Waters from Ship Surveys, 2002–2008.* Canadian Technical Report of Fisheries and Aquatic Sciences No. 2913. Nanaimo, BC: Fisheries and Oceans Canada, 2010.
Ford, John, Steven Koot, Svein Vagle, Neil Hall-Patch, and George Kamitakahara. *Passive Acoustic Monitoring of Large Whales in Offshore Waters of British Columbia.* Nanaimo, BC: Fisheries and Oceans Canada (DFO), 2010.
Francis, Daniel, and Gil Hewlett. *Operation Orca: Springer, Luna and the Struggle to Save West Coast Killer Whales.* Madeira Park, BC: Harbour, 2007.
Freccero, Carla. "Wolf, or Homo Homini Lupus." In *Arts of Living on a Damaged Planet: Ghosts and Monsters of the Anthropocene*, edited by Anna Tsing, Heather Swanson, Elaine Gan, and Nils Bubandt, 91–106. Minneapolis: University of Minnesota Press, 2017. https://www.jstor.org/stable/10.5749/j.ctt1qft070.24.
Friedner, Michele Ilana. *Sensory Futures: Deafness and Cochlear Implant Infrastructures in India.* Minneapolis: University of Minnesota Press, 2022.
Gabrys, Jennifer. *Program Earth: Environmental Sensing Technology and the Making of a Computational Planet.* Minneapolis: University of Minnesota Press, 2016.
Gabrys, Jennifer. "Programming Environments: Environmentality and Citizen Sensing in the Smart City." In *Smart Urbanism: Utopian Vision or False Dawn?*,

edited by Simon Marvin, Colin McFarlane, and Andrés Luque-Ayala, 88–107. New York: Routledge, 2015.

Gaetz, Wells. "Company Responsible for Underwater Noise Monitoring Project Hopes to Have Global Impact." *Vancouver Island CTV News*, December 13, 2020. https://vancouverisland.ctvnews.ca/company-responsible-for-underwater-noise-monitoring-project-hopes-to-have-global-impact-1.5229766.

Galison, Peter. "Trading Zone: Coordinating Action and Belief." In *The Science Studies Reader*, edited by Mario Biagioli, 137–60. New York: Routledge, 1999.

Gallagher, Michael. "Field Recording and the Sounding of Spaces." *Environment and Planning D: Society and Space* 33, no. 3 (2015): 560–76. https://doi.org/10.1177/0263775815594310.

Gallagher, Michael, Anja Kanngieser, and Jonathan Prior. "Listening Geographies: Landscape, Affect and Geotechnologies." *Progress in Human Geography* 41, no. 5 (2017): 618–37. https://doi.org/10.1177/0309132516652952.

Galloway, Alexander R., and Eugene Thacker. *The Exploit: A Theory of Networks*. Minneapolis: University of Minnesota Press, 2013.

Galois, Robert. *A Voyage to the North West Side of America: The Journals of James Colnett, 1786–89*. Vancouver: UBC Press, 2011.

Gardiner, Ray. "Railroads Built This Country." *Prince Rupert Daily News*, June 13, 1996. Prince Rupert Archives, Prince Rupert, BC.

Gassman, Martin, Lee Kindberg, Sean M. Wiggins, and John A. Hildebrand. "Underwater Noise Comparison of Pre- and Post-retrofitted Maersk G-Class Container Vessels." Report No. MPL-TM 616. University of California, San Diego, 2017.

Gautier, Ana María Ochoa. *Aurality: Listening and Knowledge in Nineteenth-Century Colombia*. Durham, NC: Duke University Press, 2015.

Gedamke, Jason, Jolie Harrison, Leila Hatch, Robyn Angliss, Jay Barlow, Catherine Berchok, Chris Caldow, et al. *Ocean Noise Strategy Roadmap*. National Oceanic and Atmospheric Administration, 2016.

George, Rose. *Ninety Percent of Everything: Inside Shipping, the Invisible Industry That Puts Clothes on Your Back, Gas in Your Car, and Food on Your Plate*. New York: Metropolitan, 2013.

Ghosh, Amitav. *The Great Derangement: Climate Change and the Unthinkable*. Berlin Family Lectures. Chicago: University of Chicago Press, 2017.

Gibson, Dan. *Solitudes—Volume 1*. Banff, AB: Digital Funding LLC, 1981.

Gieryn, Thomas F. "Boundary-Work and the Demarcation of Science from Non-science: Strains and Interests in Professional Ideologies of Scientists." *American Sociological Review* 48, no. 6 (1983): 781–95. https://doi.org/10.2307/2095325.

Gillespie, Tarleton. *Custodians of the Internet: Platforms, Content Moderation, and the Hidden Decisions That Shape Social Media*. New Haven, CT: Yale University Press, 2018.

Goeman, Mishuana. *Mark My Words: Native Women Mapping Our Nations*. Minneapolis: University of Minnesota Press, 2013.

Goonewardena, Kanishka. "The Urban Sensorium: Space, Ideology and the Aestheticization of Politics." *Antipode* 37, no. 1 (2005): 46–71.

Gopinath, Sumanth. *The Ringtone Dialectic*. Cambridge, MA: MIT Press, 2013.

Gopinath, Sumanth, and Anna Schultz. "Sentimental Remembrance and the Amusements of Forgetting in Karl and Harty's 'Kentucky.'" *Journal of the American Musicological Society* 69, no. 2 (2016): 477–524. https://doi.org/10.1525/jams.2016.69.2.477.

Gordillo, Gastón R. *Rubble: The Afterlife of Destruction*. Durham, NC: Duke University Press, 2014.

Government of Canada. "Canada's Asia-Pacific Gateway and Corridor Initiative." 2006. Accessed May 19, 2021. https://publications.gc.ca/collections/Collection/T22-131-2006E.pdf.

Government of Canada. "The Prime Minister of Canada Announces the National Oceans Protection Plan." 2016. Accessed May 19, 2021. https://pm.gc.ca/en/news/news-releases/2016/11/07/prime-minister-canada-announces-national-oceans-protection-plan.

Government of Canada. "Sgaan Kinghlas–Bowie Seamount Gin Siigee Tl'a Damaan Kinggangs Gin K'aalaagangs Marine Protected Area Management Plan 2019." Accessed August 27, 2019. https://www.dfo-mpo.gc.ca/oceans/publications/sk-b-managementplan-plangestion/page05-eng.html.

Gray, Robin R. R. "Rematriation: Ts'msyen Law, Rights of Relationality, and Protocols of Return." *Native American and Indigenous Studies* 9, no. 1 (2022): 1–27. https://doi.org/10.1353/nai.2022.0010.

Greco, Nancy. "Hear That? It's IBM's Acoustic Insights Program." Interview. *Smart Industry*, January 6, 2020. https://www.smartindustry.com/benefits-of-transformation/advanced-control/article/11296660/hear-that-its-ibms-acoustic-insights-program.

Grusin, Richard. "Radical Mediation." *Critical Inquiry* 42, no. 1 (2015): 124–48. https://doi.org/10.1086/682998.

Gumbs, Alexis Pauline. *Undrowned: Black Feminist Lessons from Marine Mammals*. Chico, CA: AK Press, 2020.

Guthman, Julie. *Wilted: Pathogens, Chemicals, and the Fragile Future of the Strawberry Industry*. Berkeley: University of California Press, 2019.

Gyibaaw. *Ancestral War Hymns*. Tape. Vancouver, BC: Ross Bay Cult Distro, 2009.

Gyibaaw. "Tour Statement." Unpublished document, 2010.

Hagen, Ross. "Musical Style, Ideology, and Mythology in Norwegian Black Metal." In *Metal Rules the Globe: Heavy Metal Music around the World*, edited by Jeremy Wallach, Harris M. Berger, and Paul D. Greene, 180–200. Durham, NC: Duke University Press, 2011.

Hagood, Mack. *Hush: Media and Sonic Self-Control*. Durham, NC: Duke University Press, 2019.

Haiven, Max. *Art after Money, Money after Art: Creative Strategies against Financialization*. London: Pluto, 2018.

Hall, Stuart. "Black Diaspora Artists in Britain: Three 'Moments' in Post-war History." *History Workshop Journal* 61, no. 1 (2006): 1–24. https://doi.org/10.1093/hwj/dbi074.

Hall, Stuart. "Gramsci's Relevance for the Study of Race and Ethnicity." *Journal of Communication Inquiry* 10, no. 2 (1986): 5–27. https://doi.org/10.1177/019685998601000202.

Hall, Stuart. "Race, Articulation, and Societies Structured in Dominance." In *Essential Essays*, vol. 1, *Foundations of Cultural Studies*, edited by David Morley. Durham, NC: Duke University Press, 2018.

Halpern, Ida. "On the Interpretation of 'Meaningless-Nonsensical Syllables' in the Music of the Pacific Northwest Indians." *Ethnomusicology* 20, no. 2 (1976): 253–71.

Halpern, Orit, and Robert Mitchell. *The Smartness Mandate*. Cambridge, MA: MIT Press, 2023.

Halseth, Greg, and Laura Ryser. "Rapid Change in Small Towns: When Social Capital Collides with Political/Bureaucratic Inertia." *Community Development* 47, no. 1 (2016): 106–21. https://doi.org/10.1080/15575330.2015.1105271.

Han, Lisa. "Precipitates of the Deep Sea: Seismic Surveys and Sonic Saturation." In *Saturation: An Elemental Politics*, 223–43. Durham, NC: Duke University Press, 2021.

Haraway, Donna. "Situated Knowledges: The Science Question in Feminism and the Privilege of Partial Perspective." In *Space, Gender, Knowledge: Feminist Readings*, edited by Linda McDowell and Joanne P. Sharp, 53–72. Abingdon, UK: Routledge, 2016.

Harris, Catriona M., Len Thomas, Dina Sadykova, Stacy L. DeRuiter, Peter L. Tyack, Brandon L. Southall, Andrew J. Read, and Patrick J. O. Miller. "The Challenges of Analyzing Behavioral Response Study Data: An Overview of the MOCHA (Multi-study OCean Acoustics Human Effects Analysis) Project." In *The Effects of Noise on Aquatic Life II*, edited by Arthur N. Popper and Anthony Hawkins, 399–407. Advances in Experimental Medicine and Biology. New York: Springer, 2016. https://doi.org/10.1007/978-1-4939-2981-8_47.

Hart, Gillian. "D/developments after the Meltdown." *Antipode* 41, no. 1 (2010): 117–41.

Harvey, David. "The Condition of Postmodernity." In *The New Social Theory Reader*, 2nd ed., edited by Steven Seidman and Jeffrey C. Alexander. New York: Routledge, 2008.

Harvey, David. "The Enigma of Capital and the Crisis This Time." In *Business as Usual: The Roots of the Global Financial Meltdown*, edited by Craig Calhoun and Georgi Derluguian, 89–112. New York: New York University Press, 2011. https://doi.org/10.18574/nyu/9780814772775.003.0005.

Harvey, David. "On the History and Present Condition of Geography: An Historical Materialist Manifesto." *Professional Geographer* 36, no. 1 (1984): 1–11. https://doi.org/10.1111/j.0033-0124.1984.00001.x.

Haskell, David George. *Sounds Wild and Broken: Sonic Marvels, Evolution's Creativity, and the Crisis of Sensory Extinction*. New York: Viking, 2022.

Heesemann, Martin, Tania L. Insua, Martin Scherwath, S. Kim Juniper, and Kate Moran. "Ocean Networks Canada: From Geohazards Research Laboratories to Smart Ocean Systems." *Oceanography* 27, no. 2 (2014): 151–53.

Hegel, G. W. F. *Aesthetics: Lectures on Fine Art.* Vol. 1. Translated by T. M. Knox. London: Oxford University Press, 1975.

Heinrich, Michael. *An Introduction to the Three Volumes of Karl Marx's Capital.* Translated by Alexander Locascio. New York: NYU Press, 2004. https://www.jstor.org/stable/j.ctt9qg6g7.

Heise, Kathy, and Lance Barrett-Lennard. "The Calm before the Storm: The Need for Baseline Acoustic Studies off the Central and North Coasts of British Columbia, Canada." *Bioacoustics* 17 (2008): 248–50. https://doi.org/10.1080/09524622.2008.9753836.

Helmreich, Stefan. *Alien Ocean: Anthropological Voyages in Microbial Seas.* Berkeley: University of California Press, 2009.

Helmreich, Stefan. *Sounding the Limits of Life: Essays in the Anthropology of Biology and Beyond.* Princeton, NJ: Princeton University Press, 2015.

Helmreich, Stefan. "Wave Theory ~ Social Theory." *Public Culture* 32, no. 2 (2020): 287–326. https://doi.org/10.1215/08992363-8090094.

Hemsworth, Katie. "'Feeling the Range': Emotional Geographies of Sound in Prisons." *Emotion, Space and Society* 20 (2016): 90–97. https://doi.org/10.1016/j.emospa.2016.05.004.

Hesse, Markus, and Jean-Paul Rodrigue. "The Transport Geography of Logistics and Freight Distribution." *Journal of Transport Geography* 12, no. 3 (2004): 171–84. https://doi.org/10.1016/j.jtrangeo.2003.12.004.

Hick, W. B. M. *Hays' Orphan: The Story of the Port of Prince Rupert.* Prince Rupert: Prince Rupert Port Authority, 2003. https://books.google.com/books?id=ECdRAAAACAAJ.

Holt, Marla M., Dawn P. Noren, and Candice K. Emmons. "Effects of Noise Levels and Call Types on the Source Levels of Killer Whale Calls." *Journal of the Acoustical Society of America* 130, no. 5 (2011): 3100–106. https://doi.org/10.1121/1.3641446.

Horn, Paul. *Haida and Paul Horn.* EP. Victoria: Sealand of the Pacific, 1972. https://www.discogs.com/release/3020564-Haida-2-Paul-Horn-Haida-Paul-Horn-EP.

Horowitz, Cara, and Michael Jasny. "Precautionary Management of Noise: Lessons from the U.S. Marine Mammal Protection Act." *Journal of International Wildlife Law and Policy* 10, no. 3–4 (2007): 225–32.

Housty, William G., Anna Noson, Gerald W. Scoville, John Boulanger, Richard M. Jeo, Chris T. Darimont, and Christopher E. Filardi. "Grizzly Bear Monitoring by the Heiltsuk People as a Crucible for First Nation Conservation Practice." *Ecology and Society* 19, no. 2 (2014). https://www.jstor.org/stable/26269572.

Howes, David. *The Sensory Studies Manifesto: Tracking the Sensorial Revolution in the Arts and Human Sciences.* Toronto: University of Toronto Press, 2022.

Hoyt, Erich. *Orca: The Whale Called Killer.* Rev. ed. Columbia, SC: Camden House, 1990.

Hracs, Brian J., Michael Seman, and Tarek E. Virani. *The Production and Consumption of Music in the Digital Age*. New York: Routledge, 2016.

Hubbs, Nadine. "Is Country Music Quintessentially American?" *American Music* 40, no. 4 (2022): 505–10. https://doi.org/10.5406/19452349.40.4.14.

Hunter, Justine. "The Sinking of the Nathan E. Stewart." *Globe and Mail*, November 3, 2016. https://www.theglobeandmail.com/news/british-columbia/inside-the-response-to-a-tug-boat-sinking-off-bcs-northerncoast/article32672711/.

Hussey, Nigel E., Steven T. Kessel, Kim Aarestrup, Steven J. Cooke, Paul D. Cowley, Aaron T. Fisk, Robert G. Harcourt, et al. "Aquatic Animal Telemetry: A Panoramic Window into the Underwater World." *Science* 348, no. 6240 (2015): 1255642. https://doi.org/10.1126/science.1255642.

Igoe, Jim. "Nature on the Move II: Contemplation Becomes Speculation." *New Proposals* 6, no. 1–2 (2013): 37–49.

Igoe, Jim. "The Spectacle of Nature in the Global Economy of Appearances: Anthropological Engagements with the Spectacular Mediations of Transnational Conservation." *Critique of Anthropology* 30, no. 4 (2010): 375–97. https://doi.org/10.1177/0308275X10372468.

Innes, Robert Alexander, and Kim Anderson, eds. *Indigenous Men and Masculinities: Legacies, Identities, Regeneration*. Winnipeg: University of Manitoba Press, 2015.

International Whaling Commission. "Report of the Standing Working Group on Environmental Concerns." 2015.

IORE. *Marine People Partnership*. Institute for Ocean Research Enterprise, October 2015. http://shipsforcanada.ca/images/uploads/Final-Report-October-2015.pdf.

Ircha, Michael C. "Serving Tomorrow's Mega-size Containerships: The Canadian Solution." *International Journal of Maritime Economics* 3, no. 3 (2001): 318–32. https://doi.org/10.1057/palgrave.ijme.9100016.

Isaksen, Gary H. "Advances to the Science of Sound and Marine Life." "Ocean Sound," special issue, *ECO: Environment Coastal and Offshore* (2019): 56–59.

Ishmael, Amelia, Zareen Price, Aspasia Stephanou, and Ben Woodard, eds. *Helvete 1: Incipit*. Santa Barbara: Punctum, 2013.

Jacobsen, Kristina M. *The Sound of Navajo Country: Music, Language, and Diné Belonging*. Chapel Hill: University of North Carolina Press, 2017.

James, Robin. *The Sonic Episteme: Acoustic Resonance, Neoliberalism, and Biopolitics*. Durham, NC: Duke University Press, 2019.

Jameson, Fredric. *Archaeologies of the Future: The Desire Called Utopia and Other Science Fictions*. London: Verso, 2005.

Jameson, Fredric. *Postmodernism, or, The Cultural Logic of Late Capitalism*. Durham, NC: Duke University Press, 1992.

Jasper, Sandra. "Sonic Refugia: Nature, Noise Abatement and Landscape Design in West Berlin." *Journal of Architecture* 23, no. 6 (2018): 936–60. https://doi.org/10.1080/13602365.2018.1505773.

Jenemann, David. *Adorno in America*. Minneapolis: University of Minnesota Press, 2007.

Johnson, Elizabeth R. "At the Limits of Species Being: Sensing the Anthropocene." *South Atlantic Quarterly* 116, no. 2 (2017): 275–92. https://doi.org/10.1215/00382876-3829401.

Johnston, Tony. *Whale Song*. New York: Putnam Juvenile, 1987.

Jones, Russ, Catherine Rigg, and Evelyn Pinkerton. "Strategies for Assertion of Conservation and Local Management Rights: A Haida Gwaii Herring Story." *Marine Policy* 80 (2017): 154–67. https://doi.org/10.1016/j.marpol.2016.09.031.

Kahn, Douglas. "Concerning the Line: Music, Noise, and Phonography." In *From Energy to Information: Representation in Science and Technology, Art, and Literature*, edited by Bruce Clarke and Linda Dalrymple Henderson, 178–94. Stanford, CA: Stanford University Press, 2002. https://doi.org/10.1515/9781503619470-016.

Kahn, Douglas. *Noise, Water, Meat: A History of Sound in the Arts*. Cambridge, MA: MIT Press, 1999.

Kahn-Harris, Keith. *Extreme Metal: Music and Culture on the Edge*. London: Berg, 2007.

Kanngieser, A. M. "Sonic Colonialities: Listening, Dispossession, and the (Re) Making of Anglo-European Nature." *Transactions of the Institute of British Geographers* (February 2023). https://doi.org/10.1111/tran.12602.

Kareiva, Peter. "Ominous Trends in Nature Recreation." *Proceedings of the National Academy of Sciences* 105, no. 8 (2008): 2757–58. https://doi.org/10.1073/pnas.0800474105.

Karjalainen, Toni-Matti, and Kimi Kärki, eds. *Modern Heavy Metal: Markets, Practices and Cultures: Conference Proceedings*. Espoo: Aalto University, 2015. https://aaltodoc.aalto.fi:443/handle/123456789/17416.

Keen, Eric. "Whales of the Rainforest: Habitat Use Strategies of Sympatric Rorqual Whales within a Fjord System." PhD diss., University of California, San Diego, 2017. https://escholarship.org/uc/item/52f602q1.

Keen, Eric M., Éadin O'Mahony, Linda M. Nichol, Brianna M. Wright, Chenoah Shine, Benjamin Hendricks, Hermann Meuter, Hussein M. Alidina, and Janie Wray. "Ship-Strike Forecast and Mitigation for Whales in Gitga'at First Nation Territory." *Endangered Species Research* 51 (2023): 31–58. https://doi.org/10.3354/esr01244.

Kelly, Lynne. *Song for a Whale*. New York: Delacorte, 2019.

Kerr, Robert. "Compression and Oppression." *CTheory*, March 28, 2013. https://journals.uvic.ca/index.php/ctheory/article/view/14795.

Keystone Environmental Ltd. Fairview Terminal Conversion Project. Environmental Screening Document. Prepared for the Prince Rupert Port Authority (PRPA), January 22, 2006. Accessed January 13, 2015. http://legacy.rupertport.com/media/fairview-terminal-phase-ii-comprehensive-study-report-en.pdf.

Kinkaid, Eden. "Re-encountering Lefebvre: Toward a Critical Phenomenology of Social Space." *Environment and Planning D: Society and Space* 38, no. 1 (2020): 167–86. https://doi.org/10.1177/0263775819854765.

Kipfer, Stefan, and Kanishka Goonewardena. "Urban Marxism and the Postcolonial Question: Henri Lefebvre and 'Colonisation.'" *Historical Materialism* 21, no. 2 (2013): 76–116. https://doi.org/10.1163/1569206X-12341297.

Klassen, Pamela E. *The Story of Radio Mind: A Missionary's Journey on Indigenous Land*. Chicago: University of Chicago Press, 2018.

Kwan, Mei-Po. "Algorithmic Geographies: Big Data, Algorithmic Uncertainty, and the Production of Geographic Knowledge." *Annals of the American Association of Geographers* 106, no. 2 (2016): 274–82.

LaBelle, B. *Acoustic Territories: Sound Culture and Everyday Life*. 2nd ed. London: Bloomsbury Academic, 2019.

LaBelle, Brandon. *Sonic Agency: Sound and Emergent Forms of Resistance*. London: Goldsmiths, 2018. https://library.oapen.org/handle/20.500.12657/63134.

Lacan, Jacques. *Ecrits*. Paris: Seuil, 1966.

Lacey, Kate. *Listening Publics: The Politics and Experience of Listening in the Media Age*. Cambridge, UK: Polity, 2013.

Lally, Nick. "Policing Sounds." *Progress in Human Geography* 47, no. 4 (2023). https://doi.org/10.1177/03091325231178029.

Landecker, Hannah. "Postindustrial Metabolism: Fat Knowledge." *Public Culture* 25, no. 3 (2013): 495–522. https://doi.org/10.1215/08992363-2144625.

Langley, Paul, and Andrew Leyshon. "Platform Capitalism: The Intermediation and Capitalisation of Digital Economic Circulation." *Finance and Society* 3, no. 1 (2017): 11–31. https://doi.org/10.2218/finsoc.v3i1.1936.

Large, R. Geddes. *Prince Rupert: A Gateway to Alaska and the Pacific*. Vancouver: Mitchell Books, 1973.

Latour, Bruno. *Facing Gaia: Eight Lectures on the New Climatic Regime*. Translated by Catherine Porter. Cambridge, UK: Polity, 2017.

Lave, Rebecca, Philip Mirowski, and Samuel Randalls. "Introduction: STS and Neoliberal Science." *Social Studies of Science* 40, no. 5 (2010): 659–75.

Lawson, J. W., and V. Lesage. "A Draft Framework to Quantify and Cumulate Risks of Impacts from Large Development Projects for Marine Mammal Populations: A Case Study Using Shipping Associated with the Mary River Iron Mine Project." DFO Can. Sci. Advis. Sec. Res. Doc. 2012/154. Fisheries and Oceans Canada, 2013.

Lefebvre, Henri. *The Production of Space*. Translated by Donald Nicholson-Smith. 1977. Malden, MA: Wiley-Blackwell, 1992.

Lefebvre, Henri. *State, Space, World*. Edited by Neil Brenner and Stuart Elden. Translated by Gerald Moore, Neil Brenner, and Stuart Elden. 1978. Minneapolis: University of Minnesota Press, 2009.

Lehman, Jessica. "A Sea of Potential: The Politics of Global Ocean Observations." *Political Geography* 55 (November 2016): 113–23. https://doi.org/10.1016/j.polgeo.2016.09.006.

Leppert, Richard. "Introduction." In *Essays on Music*, by Theodore Adorno, edited and translated by Richard Leppert, translated by Susan H. Gillespie, 1–84. Berkeley: University of California Press, 2002.

Leppert, Richard. "Music 'Pushed to the Edge of Existence' (Adorno, Listening, and the Question of Hope)." *Cultural Critique*, no. 60 (2005): 92–133.

Leppert, Richard. *The Sight of Sound: Music, Representation, and the History of the Body*. Berkeley: University of California Press, 1993.

Letham, Bryn, Andrew Martindale, Kisha Supernant, Thomas J. Brown, Jerome S. Cybulski, and Kenneth M. Ames. "Assessing the Scale and Pace of Large Shell-Bearing Site Occupation in the Prince Rupert Harbour Area, British Columbia." *Journal of Island and Coastal Archaeology* 14, no. 2 (2019): 163–97. https://doi.org/10.1080/15564894.2017.1387621.

Leuze, Mirjam, dir. *The Whale and the Raven*. Busse and Halberschmidt, Cedar Island Films, National Film Board of Canada, 2019.

Loon, Joost Van. *Risk and Technological Culture: Towards a Sociology of Virulence*. Abingdon, UK: Routledge, 2002. https://doi.org/10.4324/9780203466384.

Lough, Shannon. "Hanjin Scarlet and Crew Detained in Outer Harbour." *Today in BC*. Accessed July 29, 2023. https://www.todayinbc.com/news/hanjin-scarlet-and-crew-detained-in-outer-harbour/.

Low, Margaret, and Karena Shaw. "Indigenous Rights and Environmental Governance: Lessons from the Great Bear Rainforest." *BC Studies: The British Columbian Quarterly* 172 (2012): 9–33.

Lukes, Daniel. "Black Metal Machine: Theorizing Industrial Black Metal." *Helvete: A Journal of Black Metal Theory*, no. 1 (2013): 69–93.

Luque-Ayala, Andrés, and Simon Marvin. *Urban Operating Systems: Producing the Computational City*. Cambridge, MA: MIT Press, 2020.

Lusseau, David, David E. Bain, Rob Williams, and Jodi C. Smith. "Vessel Traffic Disrupts the Foraging Behavior of Southern Resident Killer Whales *Orcinus orca*." *Endangered Species Research* 6, no. 3 (2009): 211–21. https://doi.org/10.3354/esr00154.

Lütticken, Sven. *History in Motion: Time in the Age of the Moving Image*. Berlin: Sternberg, 2013.

Lutz, John Sutton. *Makúk: A New History of Aboriginal-White Relations*. Vancouver: UBC Press, 2009.

MacDonald, George, and Jerome Cybulski. "Introduction: The Prince Rupert Harbour Project." In *Perspectives on Northern Northwest Coast Prehistory*, edited by Jerome S. Cybulski, 1–24. Mercury Series. Ottawa: University of Ottawa Press, 2001.

MacDonald, George, and Richard Inglis. "An Overview of the North Coast Prehistory Project (1966–1980)." *BC Studies*, no. 48 (1980): 37–63.

MacFarlane, Key. "Governing the Noisy Sphere: Geographies of Noise Regulation in the US." *Environment and Planning C: Politics and Space* 38, no. 3 (2020): 539–56. https://doi.org/10.1177/2399654419872774.

Maguire, James, and Brit Ross Winthereik. "Digitalizing the State: Data Centres and the Power of Exchange." *Ethnos* 86, no. 3 (2021): 530–51. https://doi.org/10.1080/00141844.2019.1660391.

Malm, Andreas. *The Progress of This Storm: Nature and Society in a Warming World*. New York: Verso, 2018.

Mann, Geoff. "From Countersovereignty to Counterpossession?" *Historical Materialism* 24, no. 3 (2016): 45–61. https://doi.org/10.1163/1569206X-12341484.
Mann, Geoff. "Why Does Country Music Sound White? Race and the Voice of Nostalgia." *Ethnic and Racial Studies* 31, no. 1 (2008): 73–100. https://doi.org/10.1080/01419870701538893.
Markey, Sean, and Karen Heisler. "Getting a Fair Share: Regional Development in a Rapid Boom-Bust Rural Setting." *CJRS* 33, no. 3 (2010): 49–62.
Martineau, Jarrett, and Eric Ritskes. "Fugitive Indigeneity: Reclaiming the Terrain of Decolonial Struggle through Indigenous Art." *Decolonization: Indigeneity, Education and Society* 3, no. 1 (2014): I–XII. https://jps.library.utoronto.ca/index.php/des/article/view/21320.
Marx, Karl. *Economic and Philosophic Manuscripts*. Translated by Martin Milligan. Moscow: Progress, 1959.
McAllister, Ian, and Karen McAllister. *The Great Bear Rainforest: Canada's Forgotten Coast*. San Francisco: Sierra Club Books, 1998.
McCarthy, James. "Limits/Natural Limits." In *The Dictionary of Human Geography*, edited by Derek Gregory, Ron Johnston, Geraldine Pratt, Michael Watts, and Sarah Whatmore. Hoboken, NJ: John Wiley, 2009.
McCartney, Andra. "'How Am I to Listen to You?': Soundwalking, Intimacy, and Improvised Listening." In *Negotiated Moments: Improvisation, Sound, and Subjectivity*, 37–54. Durham, NC: Duke University Press, 2016.
McCauley, Robert. *Review of Documents Associated with Assessing Environmental Impacts of Underwater Noise from the Proposed Enbridge Northern Gateway Project*. Report prepared for World Wildlife Fund (WWF). Project CMST1123. World Wildlife Fund, 2012.
McCreary, Tyler. "Mining Aboriginal Success: The Politics of Difference in Continuing Education for Industry Needs." *Canadian Geographies / Géographies Canadiennes* 57, no. 3 (2013): 280–88. https://doi.org/10.1111/cag.12021.
McDonald, Mark A., John A. Hildebrand, and Sean M. Wiggins. "Increases in Deep Ocean Ambient Noise in the Northeast Pacific West of San Nicolas Island, California." *Journal of the Acoustical Society of America* 120, no. 2 (2006): 711–18. https://doi.org/10.1121/1.2216565.
McIntyre, Joan, ed. *Mind in the Waters: A Book to Celebrate the Consciousness of Whales and Dolphins*. New York: Charles Scribner's Sons, 1973.
McVay, Scott. "The Last of the Great Whales." *Scientific American* 215, no. 2 (1966): 13–21.
McWhinnie, Lauren, Leh Smallshaw, Norma Serra-Sogas, Patrick D. O'Hara, and Rosaline Canessa. "The Grand Challenges in Researching Marine Noise Pollution from Vessels: A Horizon Scan for 2017." *Frontiers in Marine Science* 4 (2017). https://www.frontiersin.org/articles/10.3389/fmars.2017.00031.
Meggs, Geoff. *Salmon: The Decline of the British Columbia Fishery*. Vancouver: Douglas and McIntyre, 1991.
Meintjes, Louise. *Sound of Africa! Making Music Zulu in a South African Studio*. Durham, NC: Duke University Press, 2003.

Menzies, Charles R. "Us and Them: The Prince Rupert Fishermen's Co-op and Organized Labour, 1931–1989." *Labour / Le Travail* 48 (fall 2001): 89–108. https://doi.org/10.2307/25149162.

Mercado, Eduardo, III, and Christina E. Perazio. "Similarities in Composition and Transformations of Songs by Humpback Whales (*Megaptera novaeangliae*) over Time and Space." *Journal of Comparative Psychology* 135, no. 1 (2021): 28–50. https://doi.org/10.1037/com0000268.

Metal Nation. "Understanding Black Metal; a Native American Point of View." July 13, 2010. https://theruraldemocrat.typepad.com/metalnation/2010/07/understanding-black-metal-a-native-american-point-of-view.html.

Miles, P. R., and C. I. Malme. *The Acoustic Environment and Noise Exposure of Humpback Whales in Glacier Bay, Alaska*. Cambridge, MA: Bolt Beranek and Newman, 1983.

Miller, Jay. *Tsimshian Culture: A Light through the Ages*. Lincoln: University of Nebraska Press, 2000.

Mills, Mara. "Deaf Jam: From Inscription to Reproduction to Information." *Social Text* 28, no. 1 (2010): 35–58. https://doi.org/10.1215/01642472-2009-059.

Moore, Dene. "B.C. Ocean Observation Project Useful for Oil Industry: Report." *Globe and Mail*, June 23, 2015. https://www.theglobeandmail.com/news/british-columbia/bc-ocean-observation-project-useful-for-oil-industry-report/article25069515/.

Moore, Jason W. *Capitalism in the Web of Life: Ecology and the Accumulation of Capital*. New York: Verso, 2015.

Morton, Alexandra. *Listening to Whales: What the Orcas Have Taught Us*. New York: Ballantine, 2004.

Morton, Timothy. "At the Edge of the Smoking Pool of Death: Wolves in the Throne Room." *Helvete: A Journal of Black Metal Theory*, no. 1 (2013). https://www.academia.edu/3253006/At_the_Edge_of_the_Smoking_Pool_of_Death_Wolves_in_the_Throne_Room.

Morton, Timothy. *Dark Ecology: For a Logic of Future Coexistence*. New York: Columbia University Press, 2016. https://doi.org/10.7312/mort17752.

Mowat, Farley. *A Whale for the Killing*. Baltimore, MD: Penguin Books, 1972.

Mowitt, John. *Percussion: Drumming, Beating, Striking*. Durham, NC: Duke University Press, 2002.

Moynihan, Michael, and Didrik Søderlind. *Lords of Chaos: The Bloody Rise of the Satanic Metal Underground*. Port Townsend, WA: Feral House, 2003.

MTV. *Pipeline Wars*. MTV Impact, September 2012. Accessed April 22, 2019. http://www.mtv.ca/shows/mtv-impact/video/season-1/mtv-news-impact-pipeline-wars/1693503/0/0.

Mudrian, Albert. "Inquisition Frontman Dagon: 'I'm Not a Nazi.'" *Decibel Magazine*, July 10, 2015. https://www.decibelmagazine.com/2015/07/10/inquisition-frontman-dagon-i-m-not-a-nazi/.

Mulvin, Dylan. "Talking It Out: An Interview with Mara Mills." *Seachange*, January 2012. https://www.academia.edu/39021691/Talking_it_Out_An_Interview_with_Mara_Mills_by_Dylan_Mulvin_.

Muñoz-Fuentes, Violeta, Chris T. Darimont, Robert K. Wayne, Paul C. Paquet, and Jennifer A. Leonard. "Ecological Factors Drive Differentiation in Wolves from British Columbia." *Journal of Biogeography* 36, no. 8 (2009): 1516–31. https://doi.org/10.1111/j.1365-2699.2008.02067.x.

Mustill, Tom. *How to Speak Whale: The Power and Wonder of Listening to Animals*. New York: Grand Central, 2022.

Myers, Natasha. "Becoming Sensor in Sentient Worlds: A More-Than-Natural History of a Black Oak Savannah." In *Between Matter and Method: Encounters in Anthropology and Art*, edited by Gretchen Bakke and Marina Peterson, 73–96. New York: Routledge, 2020.

Nadasdy, Paul. *Hunters and Bureaucrats: Power, Knowledge, and Aboriginal-State Relations in the Southwest Yukon*. Vancouver: UBC Press, 2003.

National Research Council. *Marine Mammal Populations and Ocean Noise: Determining When Noise Causes Biologically Significant Effects*. Washington, DC: National Academies Press, 2005. https://doi.org/10.17226/11147.

National Research Council. *Ocean Noise and Marine Mammals*. Washington, DC: National Academies Press, 2003. http://www.ncbi.nlm.nih.gov/books/NBK221262/.

National Research Council. *Present and Future Civil Uses of Underwater Sound*. Washington, DC: National Academies Press, 1970.

Newmark, Eddie, and Herman D. Gimbel, prods. *Ambience One (An Adventure in Environmental Sound)*. Audio Fidelity AFSD 6237, 1970. https://www.discogs.com/release/609706-No-Artist-Ambience-One-An-Adventure-In-Enviromental-Sound.

Neylan, Susan. *The Heavens Are Changing: Nineteenth-Century Protestant Missions and Tsimshian Christianity*. Vol. 31. Montreal: McGill-Queen's University Press, 2002.

Nichol, Linda M., Edward J. Gregr, Rowenna Flinn, John K. B. Ford, Riccoh Gurney, Linda Michaluk, and Allison Peacock. *British Columbia Commercial Whaling Catch Data 1908 to 1967: A Detailed Description of the B.C. Historical Whaling Database*. Canadian Technical Report of Fisheries and Aquatic Sciences 2396. Nanaimo, BC: Fisheries and Oceans Canada, Science Branch, Pacific Region, 2002

Nordhaus, Ted, and Michael Shellenberger. "The Long Death of Environmentalism." *Breakthrough Journal*, February 25, 2011. https://thebreakthrough.org/issues/energy/the-long-death-of-environmentalism.

Norgard, Tammy, Cherisse Du Preez, Jaasaljuus Yakgujanaas, Molly Clarkson, Lais Chaves, Robert Rangeley, Alessia Ciraolo, et al. "Northeast Pacific Seamount Expedition: Exploring Canada's Seamounts." *Oceanography* 32, no. 1, suppl. (March 2019): 42–43.

Norris, Pat Wastell. *Time and Tide: A History of Telegraph Cove*. Raincoast Chronicles 16. Madeira Park, BC: Harbour, 1995.

North Coast Cetacean Society. "Final Written Argument of the North Coast Cetacean Society." May 31, 2013. https://docs2.cer-rec.gc.ca/ll-eng/llisapi.dll/fetch/2000/90464/90552/384192/620327/624910/700068/960262/D152-12-1

_North_Coast_Cetacean_Society_-_NCCS_Final_argument_copy_-_A3IoZ7.pdf?nodeid=960263&vernum=-2.

Norton, Justin. "Ex-Skinhead: 'This Was Never Just about Inquisition.'" *Decibel Magazine*, July 10, 2015. https://www.decibelmagazine.com/2015/07/10/ex-skinhead-this-was-never-just-about-inquisition/.

Nowlan, L., H. Alidina, M. Ambach, L. Blight, J. Casey, and R. Powell. "WWF-Canada Submission to the Enbridge Northern Gateway Joint Review Panel." World Wildlife Fund, 2013. http://awsassets.wwf.ca/downloads/wwf_submission_to_eng_jrp_corrections_sheet_91912.pdf.

Noys, Benjamin. "'Remain True to the Earth!': Remarks on the Politics of Black Metal." In *Hideous Gnosis: Black Metal Theory Symposium*, edited by Nicola Masciandaro, 105–28. Lexington: Glossator, 2010. http://books.google.co.uk/books/about/Hideous_Gnosis.html?id=GCM2MFeJwfMC&redir_esc=y.

NRDC. "Sounding the Depths: The Rising Toll of Sonar, Shipping and Industrial Ocean Noise on Marine Life." Natural Resources Defense Council, 1999.

NRDC. "Submission of the Natural Resources Defense Council to the Enbridge Northern Gateway Project Joint Review Panel Regarding Underwater Noise Impacts from Northern Gateway Tanker Traffic." Natural Resources Defense Council, December 22, 2011.

Obee, Bruce. *Guardians of the Whales: The Quest to Study Whales in the Wild*. Anchorage: Alaska Northwest, 1992.

Ochoa Gautier, Ana María. *Aurality: Listening and Knowledge in Nineteenth-Century Colombia*. Sign, Storage, Transmission. Durham, NC: Duke University Press, 2014.

Ogden, Laura A. *Loss and Wonder at the World's End*. Durham, NC: Duke University Press, 2021.

Oldfield, Terry. *Out of the Depths (De Profundis)*. New World Company, NWCD 252, 1993. https://www.discogs.com/release/195281-Terry-Oldfield-Out-Of-The-Depths-De-Profundis.

Ommer, Rosemary E. *Coasts under Stress: Restructuring and Social-Ecological Health*. Montreal: McGill-Queen's University Press, 2007.

ONC. "Device Details: Ocean Networks Canada—Oceans 3.0." Ocean Networks Canada. Accessed August 4, 2023. https://data.oceannetworks.ca/DeviceListing?DeviceId=23482.

ONC. "Indigenous Community Engagement Plan." Ocean Networks Canada, 2014.

ONC. "March Report on the Port Metro Vancouver Underwater Listening Station." Ocean Networks Canada, 2016.

ONC. "Prince Rupert Community Observatory." Ocean Networks Canada. Accessed February 8, 2024. https://data.oceannetworks.ca/PrinceRupertPort?rotatemin=0&refreshsec=0&qpddr=L10.

ONC. "Prince Rupert—Ts'msyen Territory Community Observatory—Information for Community Members." Ocean Networks Canada, 2016. Accessed September 18, 2018. https://wwwstatic01.oceannetworks.ca/sites/default/files/pdf/Prince%20Rupert%20Information%20Package%20March%202016.pdf.

ONC. "Products and Services Deliverables Report—Transport Canada." Ocean Networks Canada, 2017.

ONC. "Strategic Plan 2030." Ocean Networks Canada, 2021. Accessed March 18, 2022. https://issuu.com/oceannetworks/docs/strategicplan_2021_issuu.

ONC. "Teacher Info." Ocean Networks Canada. Accessed August 4, 2023. https://www-static01.oceannetworks.ca/learning/ocean-sense/teacher-info.html.

Oreskes, Naomi. *Science on a Mission: How Military Funding Shaped What We Do and Don't Know about the Ocean*. Chicago: University of Chicago Press, 2022.

Palmer, Michaela, and Owain Jones. "On Breathing and Geography: Explorations of Data Sonifications of Timespace Processes with Illustrating Examples from a Tidally Dynamic Landscape (Severn Estuary, UK)." *Environment and Planning A* 46, no. 1 (2014): 222–40.

Parks, Lisa. *Cultures in Orbit: Satellites and the Televisual*. Console-ing Passions: Television and Cultural Power. Durham, NC: Duke University Press, 2005.

Pasternak, Shiri. *Grounded Authority: The Algonquins of Barriere Lake against the State*. Minneapolis: University of Minnesota Press, 2017.

Payne, Katharine, and Roger Payne. "Large Scale Changes over 19 Years in Songs of Humpback Whales in Bermuda." *Zeitschrift für Tierpsychologie* 68, no. 2 (1985): 89–114. https://doi.org/10.1111/j.1439-0310.1985.tb00118.x.

Payne, Roger. *Among Whales*. New York: Scribner, 1995.

Payne, Roger S., and Scott McVay. "Songs of Humpback Whales." *Science* 173 (1971): 585–97. https://doi.org/10.1126/science.173.3997.585.

Peters, John Durham. *Speaking into the Air: A History of the Idea of Communication*. New ed. Chicago: University of Chicago Press, 2001.

Petroleum Human Resources Council of Canada. *Labor Demand Outlook for BC's Natural Gas Industry: Petroleum Labor Market Information*. Vancouver: BC Natural Gas Workforce Strategy Committee, 2013. Accessed February 15, 2018. http://irrealty.ca/wp-content/uploads/2013/04/2013-02-21_final_bc_natural_gas_labour_demand_to_2020_report.pdf.

Peyton, Jonathan. *Unbuilt Environments: Tracing Postwar Development in Northwest British Columbia*. Nature | History | Society. Vancouver: UBC Press, 2017. https://press.uchicago.edu/ucp/books/book/distributed/U/b070043321.html.

Philippine Overseas Employment Administration. "Deployed Overseas Filipino Workers by Country/Destination (Total)." 2020. https://web.archive.org/web/20200924223605/https:/www.poea.gov.ph/ofwstat/compendium/2016-2017%20deployment%20by%20country.pdf.

Picoult, Jodi. *Songs of the Humpback Whale*. New York: Pocket Books, 2001.

Pijanowski, Bryan. *Principles of Soundscape Ecology: Discovering Our Sonic World*. Chicago: University of Chicago Press, 2024.

Pilkington, J., H. Meuter, and J. Wray. "Written Evidence Submission to the Joint Review Panel assessing the Northern Gateway Pipeline Application. North Coast Cetacean Society." Unpublished document. 2011. Accessed May 12, 2014. http://www.ceaa.gc.ca/050/documents/54935/54935E.pdf.

Pink Floyd. *Meddle*. Harvest Records, 1971. SMAS-832. https://www.discogs.com/master/20649-Pink-Floyd-Meddle.

Popova, Maria. "Nature Is Always Listening: The Science of Mushrooms, Music, and How Sound Waves Stimulate Mycelial Growth." *The Marginalian*. Accessed December 1, 2022. https://www.themarginalian.org/2022/11/13/mushrooms-sound-waves/.

Popper, Arthur N., and Anthony Hawkins. *The Effects of Noise on Aquatic Life*. New York: Springer Science and Business Media, 2012.

Prescott-Steed, David. "Frostbite on My Feet: Representations of Walking in Black Metal Visual Culture." *Helvete: A Journal of Black Metal Theory*, no. 1 (2013): 45–68.

Pugliese, Joseph. *Biopolitics of the More-Than-Human*. Durham, NC: Duke University Press, 2020.

Prince Rupert Port Authority. "Notice of Intent to Pass Whistle Cessation Resolution." 2012.

Puar, Jasbir K. *The Right to Maim: Debility, Capacity, Disability*. ANIMA: Critical Race Studies Otherwise. Durham, NC: Duke University Press, 2017.

Radovac, Lilian. "The 'War on Noise': Sound and Space in La Guardia's New York." *American Quarterly* 63, no. 3 (2011): 733–60.

Randon, Marine, Michael Dowd, and Ruth Joy. "A Real-Time Data Assimilative Forecasting System for Animal Tracking." *Ecology* 103, no. 8 (2022). https://doi.org/10.1002/ecy.3718.

Revill, George. "How Is Space Made in Sound? Spatial Mediation, Critical Phenomenology and the Political Agency of Sound." *Progress in Human Geography* 40, no. 2 (2016): 240–56. https://doi.org/10.1177/0309132515572271.

Ricketts, E. F. *Ed Ricketts from Cannery Row to Sitka, Alaska: Science, History, and Reflections along the Pacific Coast: A Compilation of Essays*. Edited by Janice M. Straley. Juneau: Shorefast Editions, 2015.

Rifkin, Mark. *Beyond Settler Time: Temporal Sovereignty and Indigenous Self-Determination*. Durham, NC: Duke University Press, 2017.

Riley, Frances. "Digby Disturbed: Tiny Communities off B.C.'s North Coast Standing Up to LNG Threat." *Northword*, November 25, 2016. https://web.archive.org/web/20170605140733/http://northword.ca/features/digby-disturbed/.

Ritts, Max, and Karen Bakker. "Conservation Acoustics: Animal Sounds, Audible Natures, Cheap Nature." *Geoforum* 124 (August 2021): 144–55. https://doi.org/10.1016/j.geoforum.2021.04.022.

Ritts, Max, Stuart H. Gage, Chris R. Picard, Ethan Dundas, and Steven Dundas. "Collaborative Research Praxis to Establish Baseline Ecoacoustics Conditions in Gitga'at Territory." *Global Ecology and Conservation* 7 (July 2016): 25–38. https://doi.org/10.1016/j.gecco.2016.04.002.

Ritts, Max, and John Shiga. "Military Cetology." *Environmental Humanities* 8, no. 2 (2016): 196–214. https://doi.org/10.1215/22011919-3664220.

Ritts, Max, and Michael Simpson. "Smart Oceans Governance: Reconfiguring Capitalist, Colonial, and Environmental Relations." *Transactions of the Institute*

of British Geographers 48, no. 2 (2023): 365–79. https://doi.org/10.1111/tran .12586.
Robinson, Dylan. *Hungry Listening: Resonant Theory for Indigenous Sound Studies.* Minneapolis: University of Minnesota Press, 2020.
Rossiter, David A., and Patricia Burke Wood. "Neoliberalism as Shape-Shifter: The Case of Aboriginal Title and the Northern Gateway Pipeline." *Society and Natural Resources* 29, no. 8 (2016): 900–915. https://doi.org/10.1080/08941920.2015 .1095378.
Roth, Christopher F. *Becoming Tsimshian: The Social Life of Names.* Seattle: University of Washington Press, 2008.
Rothenberg, David. *Thousand-Mile Song: Whale Music in a Sea of Sound.* New York: Basic Books, 2010.
Rousseau, Jean-Jacques. *The Reveries of the Solitary Walker.* Translated by Charles Butterworth. 1762. Cambridge: Hackett, 1992.
Rutherford, Stephanie. "Wolfish White Nationalisms? Lycanthropic Longing on the Alt-Right." *Journal of Intercultural Studies* 41, no. 1 (2020): 60–76. https://doi .org/10.1080/07256868.2020.1704227.
Said, Edward. *On Late Style: Music and Literature Against the Grain.* London: Bloomsbury Books, 2017.
Said, Edward. "Thoughts on Late Style." *London Review of Books,* August 5, 2004. https://www.lrb.co.uk/the-paper/v26/n15/edward-said/thoughts-on-late-style.
Saito, Kohei. *Marx in the Anthropocene.* Cambridge: Cambridge University Press, 2021.
Samuels, David William. *Putting a Song on Top of It: Expression and Identity on the San Carlos Apache Reservation.* Tucson: University of Arizona Press, 2004.
Schafer, R. Murray. *The Book of Noise.* Vancouver: Price Print, 1973.
Schafer, R. Murray. *The Tuning of the World.* New York: Random House, 1977.
Schlesinger, Angela, Marie-Noël Matthews, Zizheng Li, Jorge Quijano, and David Hannay. *Aurora LNG Acoustic Study: Modelling of Underwater Sounds from Pile Driving, Rock Socket Drilling, and LNG Carrier Berthing and Transiting.* Document 01134, Version 3.0. Victoria, BC: JASCO Applied Sciences, 2016. https://projects.eao.gov.bc.ca/api/document/58923174b637cc02bea163f1/fetch /Appendix_P_Acoustic_Modelling_Final_screening.pdf.
Scholars from the Underground. "Prince George Indigenous Band Speaks about White Supremacist Black Metal." March 27, 2013. https://web.archive.org/web /20160115175911/https://scholarsfromtheunderground.com/2013/03/17/prince -george-indigenous-band-speaks-about-white-supremacist-black-metal-bands/.
Schulze, Holger. "Das sonische Kapital: Sound in den digitalen Medien." *SPIEL* 2017, no. 2 (2017): 13–30. https://doi.org/10.3726/spiel.2017.02.02.
Schwartz, Hillel. *Making Noise: From Babel to the Big Bang and Beyond.* Princeton, NJ: Princeton University Press, 2011.
Sekula, Allan. "Fish Story: Notes on Work." In *Artists Writing/Project Proposals for Documenta 11–Platform 5: Exhibition Catalogue,* edited by Gerti Fietzek, Heike Ander, and Nadja Rottner. Berlin: Hatje Cantz, 2002.

Sevilla-Buitrago, Álvaro. "Capitalist Formations of Enclosure: Space and the Extinction of the Commons." *Antipode* 47, no. 4 (2015): 999–1020. https://doi.org/10.1111/anti.12143.

Shamelessnavelgazing. "Inquisition and Black Metal's Fascism Problem." April 28, 2014. https://shamelessnavelgazing.wordpress.com/tag/gyibaaw/.

Sheldon, Dyan. *The Whales' Song*. Des Moines, IA: Turtleback, 1997.

Shiga, John. "Sonar and the Channelization of the Ocean." In *Living Stereo: Histories and Cultures of Multichannel Sound*, edited by Paul Théberge, Kyle Devine, and Tom Everrett, 85–104. New York: Bloomsbury, 2015.

Shuvera, Ryan Ben. "Southern Sounds, Northern Voices: Unsettling Borders through Country Music." *Journal of Popular Music Studies* 30, no. 4 (2018): 177–90. https://doi.org/10.1525/jpms.2018.300412.

Siegel, Robert. *Whalesong*. Lulu.com, 2016.

Simlai, Trishant. "Negotiating the Panoptic Gaze: People, Power and Conservation Surveillance in the Corbett Tiger Reserve, India." PhD diss., University of Cambridge, 2021. https://api.repository.cam.ac.uk/server/api/core/bitstreams/f7fdbfb8-2dd0-49a2-9c32-a396bc43a6dd/content.

Simmel, Georg. "The Metropolis and Mental Life." In *Social Theory Re-wired: New Connections to Classical and Contemporary Perspectives*, 3rd ed., edited by Wesley Longhofer and Daniel Winchester. Abingdon, UK: Routledge, 2023.

Simpson, Audra. *Mohawk Interruptus: Political Life across the Borders of Settler States*. Durham, NC: Duke University Press, 2014.

Simpson, Audra. "Paths toward a Mohawk Nation: Narratives of Citizenship and Nationhood in Kahnawake." In *The Indigenous Experience: Global Perspectives*, edited by Roger C. A. Maaka and Chris Andersen, 174–88. Toronto: Canadian Scholar's Press, 2006.

Simpson, Leanne Betasamosake. *As We Have Always Done: Indigenous Freedom through Radical Resistance*. Minneapolis: University of Minnesota Press, 2017.

Smart Oceans. "Media Backgrounder: From Sensors to Decisions—When Seconds Count." 2014. https://www-static01.oceannetworks.ca/sites/default/files/pdf/Smart_Oceans_Backgrounder_Oct%202014.pdf.

Smith, Neil. *Uneven Development: Nature, Capital, and the Production of Space*. Athens: University of Georgia Press, 2014.

Sonic Acts and Hilde Methi. *Sonic Acts–Dark Ecology*. Accessed February 6, 2021. https://sonicacts.com/archive/dark-ecology.

Southall, Brandon L., Anne E. Bowles, William T. Ellison, James J. Finneran, Roger L. Gentry, Charles R. Greene Jr., David Kastak, et al. "Marine Mammal Noise Exposure Criteria: Initial Scientific Recommendations." *Aquatic Mammals* 33, no. 4 (2007): 411–521. https://csi.whoi.edu/download/file/fid/Full%20Text%20Part%20I/index-3.pdf.

Spencer, Robert. "Lateness and Modernity in Theodor Adorno." In *Late Style and Its Discontents: Essays in Art, Literature, and Music*, edited by Gordon McMullan and Sam Smiles, 220–34. Oxford: Oxford University Press, 2016.

Spitzer, Michael. *Music as Philosophy: Adorno and Beethoven's Late Style*. Bloomington: Indiana University Press, 2006.

Spong, Paul. Introduction to *Mind in the Waters: A Book to Celebrate the Consciousness of Whales and Dolphins*, edited by Joan McIntyre. New York: Charles Scribner's Sons, 1973.

Spong, Paul. "Whale Communication." In *Proceedings of the Ninth Annual Conference on Biological SONAR and Diving Mammals*. Menlo Park: Stanford Research Institute, 1972.

Springer, Simon. "A Conversation with Dr. Keith Kahn-Harris." *Treehouse of Death*, 2007. Accessed February 4, 2016. https://web.archive.org/web/20071210140012/http://www.treehouseofdeath.com/.

SQCRD. "Dodge Cove Official Community Plan." Prince Rupert, BC: Skeena-Queen Charlotte Regional District, 1990. https://www.ncrdbc.com/sites/default/files/docs/development/199_dodge_cove_ocp.pdf.

SQCRD. "Regular Board Meeting Agenda." Prince Rupert, BC: Skeena-Queen Charlotte Regional District, 2015.

SQCRD. "Regular Board Meeting Held at 344 2nd Avenue West, Prince Rupert, B.C. on Friday, November 20, 2015 at 7:00PM." Accessed March 7, 2024. https://www.ncrdbc.com/sites/default/files/uploads/meetings/agendas/11-20-2015_ag_pkg_amended.pdf.

Stalk, George, and Charles McMillan. "Seizing the Continent: Opportunities for a North American Gateway." Asia Pacific Foundation of Canada, 2013.

Steingo, Gavin, and Jim Sykes. *Remapping Sound Studies*. Durham, NC: Duke University Press, 2019.

Sterne, Jonathan. *The Audible Past: Cultural Origins of Sound Reproduction*. Durham, NC: Duke University Press, 2003.

Sterne, Jonathan. "Sonic Imaginations." In *The Sound Studies Reader*, edited by Jonathan Sterne, 1–19. London: Routledge, 2012.

Sugai, Larissa Sayuri Moreira, and Diego Llusia. "Bioacoustic Time Capsules: Using Acoustic Monitoring to Document Biodiversity." *Ecological Indicators* 99 (2019): 149–52.

Sullivan, Sian. "Beyond the Money Shot; or How Framing Nature Matters? Locating *Green* at Wildscreen." *Environmental Communication* 10, no. 6 (2016): 749–62. https://doi.org/10.1080/17524032.2016.1221839.

Sunder Rajan, Kaushik. *Lively Capital: Biotechnologies, Ethics, and Governance in Global Markets*. Durham, NC: Duke University Press, 2012.

Szabo, Victor. *Turn On, Tune In, Drift Off: Ambient Music's Psychedelic Past*. Oxford: Oxford University Press, 2023.

TallBear, Kim. "An Indigenous Reflection on Working beyond the Human/Not Human." *GLQ: A Journal of Lesbian and Gay Studies* 21, no. 2 (2015): 230–35.

Tardif, Cheryl Kaye. *Whale Song*. Rev. ed. Mississauga, ON: Kunati, 2007.

Tarnoff, Ben. "The Internet Should Be a Public Good." *Jacobin*, August 31, 2016. https://jacobin.com/2016/08/internet-public-dns-privatization-icann-netflix/.

Taylor, Sunaura. *Beasts of Burden: Animal and Disability Liberation*. New York: New Press, 2017.

Teibel, Irv. *Environments 1*. New York: Syntonic Research, SD 66001, 1969. https://www.irvteibel.com/discography/environments/.

Thompson, Emily. *The Soundscape of Modernity: Architectural Acoustics and the Culture of Listening in America, 1900–1933*. Cambridge, MA: MIT Press, 2004.

Thompson, Kim-Ly, Nikkita Reece, Nicole Robinson, Havana-Jae Fisher, Natalie C. Ban, and Chris R. Picard. "'We Monitor by Living Here': Community-Driven Actualization of a Social-Ecological Monitoring Program Based in the Knowledge of Indigenous Harvesters." *FACETS* 4, no. 1 (2019): 293–314. https://doi.org/10.1139/facets-2019-0006.

Thoreau, Henry David. *Walden; or, Life in the Woods*. London: Pan Macmillan, 2016.

Tomlinson, Gary. *The Singing of the New World: Indigenous Voice in the Era of European Contact*. Cambridge: Cambridge University Press, 2007.

Transport Canada. "The Government of Canada Releases Results of the First Five Years of the Oceans Protection Plan." Government of Canada, November 7, 2022. https://www.canada.ca/en/transport-canada/news/2022/11/the-government-of-canada-releases-results-of-the-first-five-years-of-the-oceans-protection-plan.html.

Transport Canada. "Ocean Networks Canada—Smart Oceans Contribution Program." Government of Canada. Accessed February 22, 2016. https://tc.canada.ca/en/oceans-network-canada-smart-oceans-contribution-program-0.

Truax, Barry. *Acoustic Communication*. Westport, CT: Greenwood, 2001.

Tsing, Anna Lowenhaupt. *Friction: An Ethnography of Global Connection*. Princeton, NJ: Princeton University Press, 2005.

Turkle, Sherry. *Alone Together: Why We Expect More from Technology and Less from Each Other.* New York: Basic Books, 2012.

UNCTAD. *Trade and Development: Report 2020*. United Nations Conference on Trade and Development, 2020. https://unctad.org/system/files/official-document/tdr2020_en.pdf.

University of St. Andrews. "New Tool to Assess Noise Impact on Marine Mammals." Phys.org, August 22, 2014. https://phys.org/news/2014-08-tool-noise-impact-marine-mammals.html.

Urick, Robert. *Principles of Underwater Sound*. New York: McGraw-Hill, 1983.

Vancouver Fraser Port Authority. "ECHO Program: 2020 Annual Report." 2021. https://www.portvancouver.com/wp-content/uploads/2021/04/2021-04-05-ECHO-2020-Annual-report_Final-1.pdf.

Vimalassery, Manu, Juliana Hu Pegues, and Alyosha Goldstein. "Introduction: On Colonial Unknowing." *Theory and Event* 19, no. 4 (2016). https://muse.jhu.edu/pub/1/article/633283.

Virilio, Paul. *The Original Accident*. Translated by Julie Rose. Cambridge, UK: Polity, 2007.

Visser, I. "Killer Whales in New Zealand Waters: Status and Distribution with Comments on Foraging." Tutukaka, NZ: Orca Research Trust, 2007. https://www.semanticscholar.org/paper/Killer-whales-in-New-Zealand-waters%3A-Status-and-on-Visser/7c06d50413ebcb9571236fe6ad629ec61168ffde.

Vizenor, Gerald. *Manifest Manners: Narratives on Postindian Survivance.* Lincoln: University of Nebraska Press, 1999.

Voegelin, Salomé. 2019. "Sonic Materialism: Hearing the Arche-Sonic." In *The Oxford Handbook of Sound and Imagination,* vol. 2, edited by Mark Grimshaw-Aagaard, Mads Walther-Hansen, and Martin Knakkergaard, 559–77. Oxford: Oxford University Press.

Walia, Harsha. "Really Harper, Canada Has No History of Colonialism?" *Vancouver Sun,* September 27, 2009. https://vancouversun.com/news/community-blogs/really-harper-canada-has-no-history-of-colonialism.

Weichel, Andrew. "163 Wolves Killed in Second Year of B.C.'s Controversial Cull." CTV News, May 3, 2016. https://bc.ctvnews.ca/163-wolves-killed-in-second-year-of-b-c-s-controversial-cull-1.2886672.

Weilgart, L. S. "The Impacts of Anthropogenic Ocean Noise on Cetaceans and Implications for Management." *Canadian Journal of Zoology* 85, no. 11 (November 2007): 1091–116. https://doi.org/10.1139/Z07-101.

Weilgart, Linda, Hal Whitehead, Luke Rendell, and John Calambokidis. "Signal-to-Noise: Funding Structure versus Ethics as a Solution to Conflict-of-Interest." *Marine Mammal Science* 21, no. 4 (2005): 779–81. https://doi.org/10.1111/j.1748-7692.2005.tb01265.x.

Wenz, G. M. "Ambient Noise in the Ocean: Spectra and Sources." *Journal of the Acoustical Society of America* 34, no. 12 (1962): 1936.

Western Economic Diversification Canada. "Due Diligence Report—Project 000012673." 2014.

Weyler, Rex. *Greenpeace: How a Group of Ecologists, Journalists, and Visionaries Changed the World.* New York: Rodale, 2004.

Weyler, Rex. *Song of the Whale.* Garden City, NY: Doubleday, 1986.

Wheeldon, Marianne. *Debussy's Late Style.* Bloomington: Indiana University Press, 2009.

White, Howard. Introduction to *Time and Tide: A History of Telegraph Cove,* by Pat Wastell Norris. Raincoast Chronicles 16. Vancouver: Harbour, 1995.

Whitehead, Hal. "The Cultures of Whales and Dolphins." In *Whales and Dolphins: Cognition, Culture, Conservation and Human Perceptions,* edited by Philippa Brakes and Mark Peter Simmonds. New York: Routledge, 2011.

Whitehead, Hal, and John K. B. Ford. "Consequences of Culturally-Driven Ecological Specialization: Killer Whales and Beyond." *Journal of Theoretical Biology* 456 (November 2018): 279–94. https://doi.org/10.1016/j.jtbi.2018.08.015.

Whitehead, Hal, and Luke Rendell. *The Cultural Lives of Whales and Dolphins.* Chicago: University of Chicago, 2014.

Whitehead, Hal, and Linda Weilgart. "Marine Mammal Science, the U.S. Navy and Academic Freedom." *Marine Mammal Science* 11, no. 2 (April 1995): 260–63. https://doi.org/10.1111/j.1748-7692.1995.tb00526.x.

Whitehouse, Andrew. "Listening to Birds in the Anthropocene: The Anxious Semiotics of Sound in a Human-Dominated World." *Environmental Humanities* 6, no. 1 (2015): 53–71.

Whyte, Kyle P. "Indigenous Science (Fiction) for the Anthropocene: Ancestral Dystopias and Fantasies of Climate Change Crises." *Environment and Planning E: Nature and Space* 1, no. 1–2 (2018): 224–42. https://doi.org/10.1177/2514848618777621.

Williams, Raymond. *Culture and Materialism: Selected Essays*. Radical Thinkers, no. 11. London: Verso, 2005.

Williams, Rob, David E. Bain, John K. B. Ford, and Andrew W. Trites. "Behavioural Responses of Male Killer Whales to a 'Leapfrogging' Vessel." *Journal of Cetacean Research and Management* 4, no. 3 (2002): 305–10. https://doi.org/10.47536/jcrm.v4i3.844.

Williams, Rob, Christine Erbe, Erin Ashe, Amber Beerman, and Jodi Smith. "Severity of Killer Whale Behavioral Responses to Ship Noise: A Dose-Response Study." *Marine Pollution Bulletin* 79, no. 1 (2014): 254–60. https://doi.org/10.1016/j.marpolbul.2013.12.004.

Williams, Rob, Christine Erbe, Erin Ashe, and Christopher W. Clark. "Quiet(er) Marine Protected Areas." *Marine Pollution Bulletin* 100, no. 1 (2015): 154–61. https://doi.org/10.1016/j.marpolbul.2015.09.012.

Wilson, Gary, and Tracy Summerville. "Transformation, Transportation or Speculation? The Prince Rupert Container Port and Its Impact on Northern British Columbia." *Canadian Political Science Review* 2, no. 4 (2008): 26–39.

Wilson, Jacqueline. *The Longest Whale Song*. Melbourne: Bolinda Audio, 2015.

Winter, Paul. *Common Ground*. Los Angeles: A&M Records, 1977. https://www.allmusic.com/album/common-ground-mw0000196373.

World Conservation Congress. "Undersea Noise Pollution." RESWCC3.068. Third Session, Bangkok, Thailand, November 17–25, 2004. https://awionline.org/sites/default/files/uploads/documents/IUCN_RES053-1238105850-10131.pdf.

Wray, Janie, and Hermann Meuter. "Project Cetacea Lab." Save Our Seas Foundation. Accessed July 28, 2023. https://saveourseas.com/project/cetacealab/.

WWF-Canada Blog. "Cetacea Lab Archives." Accessed July 28, 2023. https://blog.wwf.ca/blog/tag/cetacea-lab/.

Zelko, Frank. *Make It a Green Peace! The Rise of Countercultural Environmentalism*. Oxford: Oxford University Press, 2013.

Zimmer, Walter M. X. *Passive Acoustic Monitoring of Cetaceans*. Reprint, Cambridge: Cambridge University Press, 2014.

Zylinska, Joanna. "Hydromedia: From Water Literacy to the Ethics of Saturation." In *Saturation: An Elemental Politics*, edited by Melody Jue and Rafico Ruiz, 45–70. Durham, NC: Duke University Press, 2021.

INDEX

Note: Page numbers in italics refer to illustrations.